NUMBER
FREAK

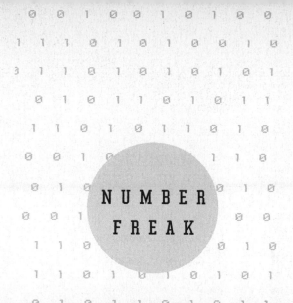

NUMBER FREAK

FROM 1 TO 200
THE HIDDEN LANGUAGE OF NUMBERS REVEALED

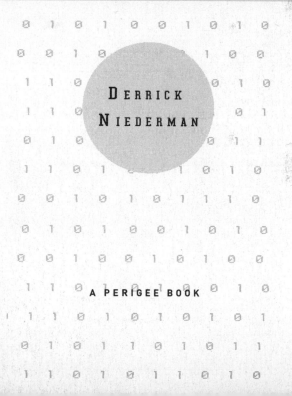

DERRICK NIEDERMAN

A PERIGEE BOOK

A PERIGEE BOOK
Published by the Penguin Group
Penguin Group (USA) Inc.
375 Hudson Street, New York, New York 10014, USA
Penguin Group (Canada), 90 Eglinton Avenue East, Suite 700, Toronto, Ontario M4P 2Y3, Canada
(a division of Pearson Penguin Canada Inc.)
Penguin Books Ltd., 80 Strand, London WC2R 0RL, England
Penguin Group Ireland, 25 St. Stephen's Green, Dublin 2, Ireland (a division of Penguin Books Ltd.)
Penguin Group (Australia), 250 Camberwell Road, Camberwell, Victoria 3124, Australia
(a division of Pearson Australia Group Pty. Ltd.)
Penguin Books India Pvt. Ltd., 11 Community Centre, Panchsheel Park, New Delhi-110 017, India
Penguin Group (NZ), 67 Apollo Drive, Rosedale, North Shore 0632, New Zealand
(a division of Pearson New Zealand Ltd.)
Penguin Books (South Africa) (Pty.) Ltd., 24 Sturdee Avenue, Rosebank, Johannesburg 2196, South Africa

Penguin Books Ltd., Registered Offices: 80 Strand, London WC2R 0RL, England

While the author has made every effort to provide accurate telephone numbers and Internet addresses at the time of publication, neither the publisher nor the author assumes any responsibility for errors, or for changes that occur after publication. Further, the publisher does not have any control over and does not assume any responsibility for author or third-party websites or their content.

First edition: August 2009

Library of Congress Cataloging-in-Publication Data

Niederman, Derrick.
    Number freak : from 1 to 200—the hidden language of numbers revealed / Derrick Niederman.—1st ed.
        p. cm.
    "A Perigee book."
    ISBN 978-0-399-53459-1
    1. Mathematics—Miscellanea. I. Title.
    QA99.N54 2009
    510—dc22   2009016032

PRINTED IN THE UNITED STATES OF AMERICA

10   9   8   7   6   5   4   3   2   1

Most Perigee books are available at special quantity discounts for bulk purchases for sales promotions, premiums, fund-raising, or educational use. Special books, or book excerpts, can also be created to fit specific needs. For details, write: Special Markets, Penguin Group (USA) Inc., 375 Hudson Street, New York, New York 10014.

NUMBER
FREAK

"NUMBERS make them number."

Those were the words of Eugene T. Maleska, former crossword editor of the *New York Times*. The year was 1981, and he had just accepted one of my efforts for publication. In doing so he inquired about my lot in life, and I replied that I was a graduate student in mathematics (at the moment making crossword puzzles instead of working on my thesis, but that's another story). He replied in turn that most word people aren't interested in mathematics, letting the quote above speak for itself.

Gene has long since left this world, but in a way this book is for people just like him—intellectually curious people intent on learning something new every day, but for whom numbers remain something of a mystery. People who have heard of a prime number but might not be able to define what it means.

As it happens, as I began writing this book, Gene's successor Will Shortz titled one of my later puzzles in a way that illustrates what this book is all about. The puzzle in question appeared in the *Times* in August 2006, and involved names and expressions such as H. L. Mencken, IQ test, MX missile, and C. S. Lewis. Will gave the puzzle the title $13 \times 2 = 26$, hoping to have dropped a big hint. The idea was that once solvers reminded themselves that the English alphabet contained 26 letters, they had a big head start in deciphering the puzzle's theme—13 entries, each beginning with a pair of letters as per the examples above, altogether representing every letter in the alphabet once and only once.

And that's what this book is about. Go to the *n*th page and you'll find out everything about the number *n* you ever wanted to know—its arithmetic, its geometry, and even its appearances in popular culture. We will discover that numbers have an individuality about them that you'd never see from afar. Just because 16 and 17 are close together, for example, doesn't mean they act the same. One is a perfect square, being 4×4, the other is prime, having no factors other than itself and 1. Sixteen is a wonderful number for a weekend tennis tournament, while seventeen stinks in that regard but excels at others. How many people are aware that there are precisely 17 symmetric wallpaper patterns?

I ended up tackling every number from 1 to 200, winnowing the discussions as I got to triple digits. I found that some numbers had enough going for them to be their own book, while others presented a struggle to find anything at all: 138, anybody? But I ended up being amazed at just how many numbers had a story to tell if you were willing to dig deep enough.

Now for a few true confessions. First, even though this book *feels* comprehensive, there are plenty of number properties that didn't quite make the cut for the simple reason that there had to be a cut. I don't think I mentioned that there are 13 witches in a coven, or that 200 is a common cutoff when assessing cholesterol readings. Sorry. And I could have written an entire book using only sports numbers, so you can imagine how many sports-related entries were moved aside to make room for others. Religious or otherwise sacred numbers could have made yet another separate book, again one that I chose not to write. This is a book about numbers, not a book about numerology, and there's a big difference. Yes, I dabbled in some numerological notions. I even observed that numbers such as 37 have built cult followings for their supposed mysticism. Even if I didn't share their particular zeal, I at least tried to show what the fuss is all about.

I also failed to give adequate explanations for expressions such as "the whole nine yards" or "23-skidoo," but in this case please don't blame me. In 95% of these cases, there's no single origin, but rather a list of theories of origin. I gave it my best shot with "86," as in "to jettison," but I found that it's tough to write when you're busy making disclaimers and caveats, so I left many if not most such numerical expressions alone. Frankly I was surprised at just how many number-related expressions had no definitive roots.

This book is stuffed with trivia, but it's also stuffed with the history of mathematics—and don't you dare equate the two! As tours to mathematical history go, this is about the most zigzagging one you can imagine. You will be in 1800 on one page and back to 200 BC on the next. But before you're done you'll encounter all the greats, from Euclid to Euler and on to modern day standouts of whom you may have never heard.

Part of me worries that I have occasionally done these great scholars a disservice. After all, there's a lot of just plain fun and silly stuff within the book, often rubbing right up against some important mathematics. Even worse, I won't necessarily tell you in advance which is which! Sometimes there is in fact a fine line between what we call "recreational mathematics" and areas in mathematics that have a wide range of useful applications. But the point is that most of the great names in the field have dabbled in both, because their curiosity wouldn't let them do otherwise. This book is ultimately about developing your mathematical mind, and there's more than one way to do it. If some of the basic mathematical terms remain foreign to you, I whipped up a glossary that should help keep you in the game.

And if you want to read the book with an eye for popular culture only, that's fine. The seven wonders of the ancient world are every bit as "sevenish" as the seven bridges of Konigsberg, and you don't need any graph theory to understand them. And you don't need me to tell you that there are seven days in a week, but I tried not to forget those everyday items as well.

The nature of this project made it a bottomless pit, but I'm almost sad that it had to come to an end. You, however, are just beginning, and I wish you well along your numerical journey.

—Derrick Niederman
 Needham, MA

1

THE number 1 is both a logical and a lousy way to start this book. Logical because 1 comes first, and its omission would seem absurd. But also lousy, because this book is about special properties of whole numbers, and the number 1 just has too many special properties for its own good.

For starters, 1 is the "multiplicative identity"—if you multiply any number by 1, that number is left unchanged. In particular, 1 is its own square and cube, and in general 1 to the nth power equals 1 for any n. A probability of 1 is the same thing as certainty, and 1 is also the maximum value attained by the basic trigonometric functions sine and cosine. The number 1 is also a trivial solution to many equations that mathematicians study, so much so that it pretty much has to be exempted: Not only is it a perfect square and a perfect cube, it's a triangular, pentagonal, and hexagonal number, and so on. You see the problem?

The idea that the number 1 is everywhere has a specific foundation in the form of Benford's Law. Introduced in 1938 by physicist Frank Benford, the idea is that the digits found in a variety of naturally occurring datasets (molecular weights, population sizes, digits on the front pages of newspapers, to name a few of the thousands of sets studied by Benford) are not uniformly distributed. In particular, the leading digit of such numbers is 1 about three-tenths of the time, far above the one-in-nine frequency you might expect. (Sorry, but zeroes don't count in this context.)

The progenitor to Benford's Law was astronomer Simon Newcomb's 1881 discovery that within a book of logarithm tables, the pages containing logarithms beginning with 1 were much more likely to be worn around the edges. More recently, Benford's Law has been applied in the detection of tax and accounting fraud. The basic principle is that people who make up numbers are not familiar with Benford's Law, and the bogus datasets they generate stand out because they have far too few 1's in them.

▼

THE Greek philosopher Parmenides (fifth century BC) was of the view that "all is one." While I can't say I understand exactly what he meant, apparently it was Parmenides who inspired Zeno (which can be spelled with either a Z or an X, followed by *one* backward) to come up with his now famous set of paradoxes. Perhaps the best known of these paradoxes, known in the trade as Dichotomy, asserts that before you can ever reach a destination, you have to go halfway first. No problem there, except that from the halfway point you have to go halfway again, and so on, the paradox being that as long as you do so you'll never quite get to your destination. This line of reasoning confounded the thinkers of Zeno's day, but by the time Newton and Leibniz drafted the concepts of calculus in the late 1600s, the notion that an infinite series of numbers could converge to a finite number was no longer paradoxical. The specific series at hand is represented by the equation

$$1 = \frac{1}{2} + \frac{1}{4} + \frac{1}{8} + \frac{1}{16} + \frac{1}{32} + \ldots, \text{ or more compactly as } 1 = \sum_{n=1}^{\infty} \frac{1}{2^n}$$

▼

A more advanced type of paradox involving the number 1 came from the great French mathematician Henri Lebesgue (1875–1941). His Lebesgue measure (pronounced *le-bayg*, by the way) provided a means for measuring various subsets of Euclidean space. A simple example would be the interval [0,1] (namely, all real numbers between 0 and 1), which has a Lebesgue measure of 1. No problem so far, but it jars our intuition when we hear that the set of irrational numbers between 0 and 1 also has measure 1. In other words, the entire set of proper fractions, infinite though it may be, has a measure of zero, a stunning reminder that not all infinities are the same.

Lebesgue was apparently the last notable mathematician to think of 1 as a prime number. Yes, the late astronomer Carl Sagan included 1 as a prime in his 1985 book *Contact*, but nowadays most mathematicians would be more swayed by the views of longtime UCLA professor and Einstein protégé Ernst Gabor Straus, to whom is attributed the following quotation: "The primes are the building bricks of arithmetic, and 1 is just not a brick!" In that spirit, 1 is the only number in this volume that will not be designated at the top of its page as being either prime or composite; in the latter case, we will of course be kind enough to provide the factorization.

IN the world of music, *A Chorus Line* highlighted one as a singular sensation, but Harry Nilsson had already warned us that it is also the loneliest number. Three Dog Night's 1969 version of the Nilsson song "One" was that band's first gold record, although it did not reach number 1 on the Billboard charts.

IN chess, the number 1 is used to indicate victory, so a line-by-line recap of a chess game that ends with 1–0 indicates a victory by white, while $\frac{1}{2} - \frac{1}{2}$ indicates a draw and 0–1 indicates a black victory.

FINALLY, the phrase "We're number one" is by now a rather tedious refrain, especially when it comes from the other side of a football stadium, but the expression isn't about to be replaced, such is the power of one. Along those lines, it must be said that the expression "public enemy number one" simply isn't what it was in the days of Al Capone. Nor is the "identity parade," known to Capone and even law-abiding Americans as the "police lineup," whose popularity has fallen with the advent of computer databases and DNA technology. But even when lineups were in vogue, many police sta-

tions started their lineups with number 2, simply because the mighty number 1 was chosen disproportionately often, no matter where the guilty party happened to be standing. Frank Benford might have predicted as much.

2 [prime]

THE number 2 is the only even prime number. It is also the only prime number that lacks an *e* in its name, but that's because every odd number has an *e*.

▼

THE square root of 2, written $\sqrt{2}$, was the first number shown to be irrational, meaning that it cannot be written as the quotient of two whole numbers. The discovery of this fact is credited to Hippasus of Metapontum (circa 500 BC), and its proof is surprisingly easy. Just assume that $\sqrt{2} = \frac{p}{q}$, with p and q positive integers with no common factors. If we square both sides we get $2 = \frac{p^2}{q^2}$, or $2q^2 = p^2$. But it's not possible for one perfect square to be twice another. Specifically, note that p can't be odd in the above equation, because if it were, the right-hand side would be odd, whereas the left-hand side is clearly even. But if p is even it has a factor of 2, and there p^2 would be divisible by 4. But since $p^2 = 2q^2$, that would mean q is even, violating our assumption that p and q have no common factor. Hence the equation $2q^2 = p^2$ cannot ever hold.

Okay, maybe that wasn't as easy as you would have liked, but don't be jealous of Hippasus. Apparently the notion of irrational numbers was an affront to his fellow Pythagoreans, a group that thought the world revolved around nice simple ratios. Hippasus made his discovery while at sea, or so the legend goes, whereupon his shipmates threw him overboard. Nowadays we are more tolerant of mathematical discovery, although it should be said that those who demonstrate an early aptitude for mathematics are often the object of derision and ridicule in grade school. The irony of the Hippasus tale is that the set of irrational numbers turns out to be a higher-order infinity than the set of rational numbers, a fundamental fact that neither he nor his tormentors lived to discover.

▼

THE prefix *bi-* means two, as in bicycle, binoculars, biplane, bisect, and binary. Curiously, the word *biscuit* also applies, as it derives from the Italian biscotti, which literally means "twice cooked."

▼

IN geometry, the number 2 shows up in the famous saying "two points determine a line." This statement is emphatically true in planar geometry, but frequently abused in the real world, where two data points may be insufficient to determine an actual trend.

▼

IT is well-known that if a right triangle has legs a and b and hypotenuse c, then $a^2 + b^2 = c^2$. This equation, known to one and all as the Pythagorean Theorem, may not even be the most recognized equation involving c^2 (that title probably belongs to $E = mc^2$), but it is still way up there on the list of recognizable mathematical equations. And it's a mite easier to prove than Einstein's trademark equation, as the following diagram shows:

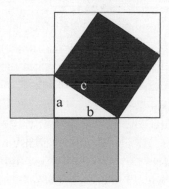

The largest square measures $a + b$ on each side, so its area equals $(a + b)^2$. But this square also consists of the square with side c coupled with four triangles each with area $\frac{ab}{2}$. Putting everything together, $(a + b)^2 = c^2 + 4(\frac{ab}{2})$. Multiplying out the left-hand side, $a^2 + 2ab + b^2 = c^2 + 2ab$, so $a^2 + b^2 = c^2$. In particular, if a and b both equal 1, c equals the square root of 2, and the exploration of what c could possibly be is apparently what got Hippasus his ticket to Davy Jones's locker. (It was fourth-century BC geometer Theaetetus who is credited with the more general result that the square root of any integer that is not a perfect square must be an irrational number.)

Note that the above proof required that the angle between a and b was a right angle, because otherwise the areas of the white triangles would have been more complicated to calculate. You should know that the Pythagorean Theorem works the other way around—if the legs of a triangle are such that $a^2 + b^2 = c^2$, then the triangle in question is a right triangle. That much was common knowledge in the time of Euclid (300 BC). Since Euclid's day there have been hundreds of published proofs of the Pythagorean Theorem. One of the most notable was devised in 1876 by James A. Garfield, who became president of the United States five years later. (Alas, nowadays anyone with such mathematical curiosity and competence would probably have it used against him in a national election.)

▼

THE Pythagorean Theorem generalizes to three dimensions, so if a, b, and c are the three sides of a rectangular prism (box) and d is its diagonal, you have the equation $a^2 + b^2 + c^2 = d^2$. And it even generalizes in two dimensions to triangles other than right triangles, yielding the law of cosines.

But the one thing you can't do to the Pythagorean Theorem is to take the 2 out of it. The celebrated Fermat's Last Theorem, proved by Princeton's Andrew Wiles in the mid-1990s after a 300-year wild-goose chase, declares that the equation $x^n + y^n = z^n$ has no solutions for any positive integer n other than 2. (Okay, I lied. That equation has infinitely many solutions when $n = 1$, proving the point made in **1** that 1 can be a nuisance, because so many mathematical truths "accidentally" include 1 as a completely trivial case.)

▼

FOR a final appearance of the number 2 in a geometric context, it shows up in a formula derived by the great Swiss mathematician Leonhard Euler (1707–1783), who noticed that in a three-dimensional shape known as a polyhedron, there is a fixed relationship between the number of edges, faces, and vertices.

To take a simple example, in the cube below there are 12 edges, 8 vertices, and 6 faces, and $8 + 6 - 12 = 2$.

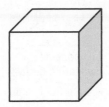

Euler showed that this formula holds for any polyhedron. Specifically, if the number of edges, vertices, and faces are denoted by e, v, and f, respectively, then $v + f - e = 2$.

(No, the e you see there isn't the e from a calculus course, though it is perhaps worth pointing out that the calculus e is named for Euler, whose extraordinary work will appear many times before this book is through.)

3 [prime]

THE "Three Legs of Man" has been the official symbol of the Isle of Man since the thirteenth century. A possible inspiration was the Sicilian flag, which features the head of Medusa surrounded by three legs. Many European flags, notably the French and Italian flags, consist of three vertical stripes.

JUST as two points determine a line, any three points that aren't in a straight line determine a plane. In particular, those three points can be joined to form a triangle, of which there are a few basic types:

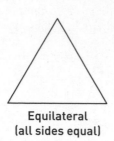
Equilateral
(all sides equal)

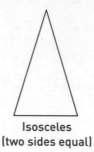
Isosceles
(two sides equal)

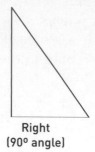
Right
(90° angle)

THREES IN GEOMETRY

IF three congruent circles intersect in a single point (the point P at right), the other three points of intersection must lie on a circle congruent to the first three. This result is known as Johnson's Theorem, having been discovered by Roger Johnson in 1913.

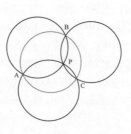

IF three equilateral triangles of different sizes are joined point-to-point (shown at right), their centers will form the vertices of a fourth equilateral triangle. This result is actually known as Napoleon's Theorem, in honor of its discoverer, an amateur mathematician named Napoleon Bonaparte. Napoleon's combination of intellect and, eventually, power, led the great astronomer and mathematician Pierre-Simon Laplace to dedicate his groundbreaking *Celestial Mechanics* to him. As legend has it, Napoleon thanked Laplace for this dedication and observed that the manuscript had

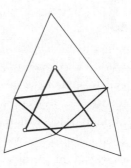

made no mention of God. To which Laplace replied, "Sire, I found I had no need of that hypothesis."

▼

IT is possible for a right triangle to be isosceles, but a right triangle can never be equilateral, as all angles of an equilateral triangle must be 60 degrees. A non-right triangle with three unequal sides is called scalene.

▼

THERE is a nice relationship that unites a perfectly symmetric equilateral triangle (above left) with a perfectly asymmetric scalene triangle (above right). The figure between them looks quite a bit like a Mercedes logo, but for our purposes it is just a three-pointed star whose endpoints form an equilateral triangle. Now imagine taking that logo and placing it inside the scalene triangle such that each point of the logo is pointed directly at one of the vertices of the triangle—if it helps to envision yourself driving a Mercedes, so be it. Once you find this point, draw lines to the three vertices as in the right-hand diagram. Whether you realized it or not, you have created three line segments such that the angle between any two of them is precisely 120 degrees (always possible unless the triangle had an angle exceeding 120 degrees in the first place). That magical point in the interior of the triangle is known as the "Fermat point" and has the property that the sum of the distances to the three vertices is the smallest possible. Finding points such as these is of interest today if you're laying cable or constructing paths on semiconductors, but back in the 1600s it was a challenge given by Pierre de Fermat to Evangelista Torricelli. Torricelli apparently solved the problem and had time left over to invent the barometer.

▼

PERHAPS the best-known triangle of all is of infinite length, and that would be Pascal's Triangle, the first few rows of which are seen below. The sides of the triangle consist of 1's, while every number in the middle is obtained by adding the two numbers to the above-left and above-right.

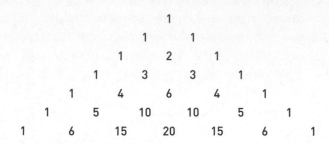

While simple in its construction, Pascal's Triangle contains a host of important patterns and rules. For example, in the bottom row above (usually referred to as the sixth row, with the single "1" on top treated as row zero), the third number from the left (15) is the number of ways of choosing a subset of three objects from an original set of 6 (6 ways of choosing the first one, multiplied by 5 ways of choosing the second, multiplied by 4 ways of choosing the third, divided by 6 ways of arranging the three objects you chose). In general, the kth entry of the nth row of Pascal's Triangle equals $\frac{n!}{k!\,(n-k)!}$ where ! is the factorial function (n factorial is the product of all positive integers less than or equal to n).

An alternative interpretation of the triangle can be seen by calculating an expression such as $(a + b)^6$, which equals $a^6 + 6a^5b + 15a^4b^2 + 20\,a^3b^3 + 15\,a^2b^4 + 6\,ab^5 + b^6$. Suddenly the sixth row of Pascal's Triangle emerges in the form of the numbers in front of the various expressions involving powers of a and b (more compactly, but more off-puttingly, known as "binomial coefficients").

▼

ALTHOUGH much of mathematics is universal, the name of Pascal's Triangle is not, as it was studied in many places in the world long before the arrival of Blaise Pascal in the 1600s. In China, Pascal's Triangle is called Yang

Hui's Triangle after the thirteenth-century Chinese mathematician who studied it. In Iran it is known as Khayyam's Triangle, while in Italy it is known as Tartaglia's Triangle. All of which raises the question of why in the world we call it Pascal's Triangle, and the answer is that a fellow named Pierre Raymond de Montmort decided to name the triangle in Pascal's honor sometime in the early eighteenth century. Here's a relatively new puzzle arising from the triangle: Suppose that after creating any particular row of Pascal's Triangle, you calculated the ratio of odd numbers to even numbers. If you kept going, that ratio would approach a fixed limit. What is that limit? (See Answers.)

▼

THE expression "Two's company, three's a crowd" applies in all sorts of situations. To take a relatively benign example, while many games are designed for one-on-one play (chess and backgammon come quickly to mind), and while many more modern board games work perfectly well with multiple players (Sorry, Clue, Monopoly), it's rare for board games to be designed specifically for *three* players. Yet many TV games, including such stalwarts as *Jeopardy!* and *Wheel of Fortune*, are designed to accommodate three participants. And *Let's Make a Deal* often made contestants choose among curtains numbered 1, 2, and 3, a feature that eventually placed emcee Monty Hall in the middle of a mathematical paradox.

What became known as the Monty Hall Problem was actually formulated by Martin Gardner as the Three Prisoner Problem as far back as 1959. In the modern version, a contestant on *Let's Make a Deal* chooses one of the three curtains in hopes that a car—as opposed to, say, a goat—was behind it. Let's say that the contestant chooses curtain number 1. In the workings of the problem, though not in *Let's Make a Deal* itself, Monty Hall then opens one of the other two curtains, revealing a goat. The question is, assuming that curtain number 2 has now been revealed (and that Monty Hall is not being mischievous), does it make sense to switch your original guess from number 1 to number 3, or should you stay put?

This problem was posed by Marilyn vos Savant in *Parade* magazine in 1990, and the volume of reader mail was extraordinary. Thousands of respondents simply couldn't accept her claim that the contestant should indeed switch her choice from curtain number 1 to curtain number 3. Many prominent mathematicians apparently entered the debate—on the wrong side.

While it is counterintuitive to think that switching curtains improves your chances—after all, the only thing Monty Hall did was to reveal a curtain that did not have the car behind it, and you already knew that such a curtain existed—there's a way of seeing through the paradox. The key is to note that Hall's actions did not change the likelihood that curtain number 1 held the car; its chances were and remained one in three. But now that curtain number 2 has been ruled out altogether, the likelihood of the car being behind curtain number 3 must have risen from $\frac{1}{3}$ to $\frac{2}{3}$.

▼

A Koch snowflake is made by starting with an equilateral triangle, removing the middle third of each side, building three new equilateral triangles to fill in the missing thirds, and continuing the process. As the number of iterations grows, the perimeter of the snowflake grows without bound, but the area never exceeds $\frac{8}{5}$ the area of the original triangle. This phenomenon plays out in the real world in the form of coastlines. For example, the coastline of Alaska, full of little jigs and jags, is 5,580 miles, almost as great as the 6,053 miles of coastline of the 48 contiguous states combined.

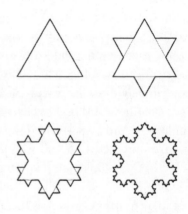

The Koch snowflake is an example of a fractal—a shape that can be split into smaller parts that mimic the whole. The father of fractals, Benoit Mandelbrot, would argue that the length of a coastline is in fact unbounded, limited only by the amount of natural detail that you wish to incorporate. Fractal-like objects occur frequently in nature and are as varied as mountain ranges, river networks, and ferns.

FAMOUS THREES

THE Big Three at Yalta were Churchill, FDR, and Stalin. A generation earlier, the Big Three would have been Lloyd George, Wilson, and Clemenceau at Versailles.

In American business, the Big Three either means the television networks ABC, CBS, and NBC or the automakers General Motors, Ford, and Chrysler. The term is applied the world over in various forms, from American colleges, as in Harvard, Yale, and Princeton, to Portuguese sports clubs, as in the *Os Três Grandes* SL Benfica, FC Porto, and Sporting CP.

THE DIVINE COMEDY

DANTE'S masterpiece is divided into three segments called *canticas*: *Inferno*, *Purgatorio*, and *Paradiso*. The three-act structure is standard in theatrical productions because it creates a natural sequence of setup, confrontation, and resolution. This same sequence appears even in four-act plays such as Arthur Miller's *The Crucible* and five-act plays such as Shakespeare's *Macbeth*.

GAUL

"ALL Gaul is divided into three parts." Julius Caesar's famous statement on Gaul was not always borne out by history: Gaul was home to many different cultural subdivisions before the fourth century AD, when the emergence of the Franks gave the country its current name and boundaries. Along these lines, the French Revolution was also characterized by *Liberte, Egalite, Fraternite*, a troika that to some is embodied by the French *tricolore* of red, white, and blue.

· · ·

4 $\left[2^2 \right]$

THE number 4 can be created by taking two to the second power, as above, or by multiplying two 2's, which boils down to the same thing, or by adding two 2's, which boils down to the same thing as that.

▼

COMBINING two 2's to make four is something of a quintessential truth: In George Orwell's *Nineteen Eighty-Four*, Winston Smith of the Ministry of Truth proclaimed, "Freedom is the freedom to say that two plus two make four. If that is granted, all else follows." Fyodor Dostoevsky preferred the multiplicative route, as in the following passage from his 1864 novel *Notes from Underground*: "Good heavens, gentlemen, what sort of free will is left when we come to tabulation and arithmetic, when it will all be a case of twice two make four? Twice two makes four without my will. As if free will meant that!"

▼

AS marvelous as the equation $4 = 2 + 2 = 2 \times 2 = 2^2$ is, it sometimes makes patterns more difficult to figure out. To wit, consider that there are precisely four subsets of a set of two elements, which we'll denote by {A,B}: Those subsets are {A}, {B}, {A,B} (the set itself), and Ø (the empty set). But what is the formula for the number of subsets of a set with n elements? Is it $2n$? Or maybe n^2? Anyone for 2^n? n^n? All those formulas work in the case where $n = 2$. As this is not *Suspense Theater*, let me reveal that the correct answer is 2^n: The principle here is one of binary inclusion/exclusion—each of the n elements is either in a particular subset or it isn't, and those two choices are multiplied by one another n times to produce the final result of 2^n.

▼

4 is the only number in the English language that contains the same number of letters as its name. But there's more going on here. Choose any number whatsoever and count the number of letters in its name to get a new number. Count the number of letters in the new number, and keep going. No matter what number you started with, you'll eventually end up with four. The proof is actually a little easier than you might think. Care to take a stab at it? (See Answers.)

▼

THERE are said to be four fundamental forces in nature—gravity, electromagnetism, and weak and strong nuclear forces—so it is fitting that a whole bunch of things are divided into fourths:

Directions/points	north	south	east	west
Cards/suits	spades	hearts	diamonds	clubs
Orchestra/sections	brass	woodwind	percussion	string
Year/seasons	spring	summer	fall	winter

No doubt you can come up with a few quartets of your own, but you should be aware that the concept of groups of four dates back to the followers of Pythagoras, aptly named the Pythagoreans. We will encounter this quirky group a bit later on, but for now let's just say that they were apparently aware of the groupings below as far back as the fifth century BC:

Numbers	1	2	3	4
Magnitudes	point	line	surface	solid
Elements	fire	air	water	earth
Figures	pyramid	octahedron	icosahedron	cube
Living Things	seed	growth in length	in breadth	in thickness
Societies	man	village	city	nation
Faculties	reason	knowledge	opinion	sensation
Seasons	spring	summer	autumn	winter
Stages of life	infancy	youth	adulthood	old age

▼

SPEAKING of groups of four, we owe another one to the Pythagoreans—the division of mathematics into four groups. The tree branches down as follows:

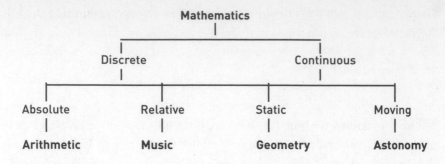

The base of the tree is the famous *Quadrivium*, the four subjects needed for a bachelor's degree in the Middle Ages.

▼

THE French mathematician and poet Claude-Gaspar Bachet de Méziriac (1581–1638) made a famous conjecture about the number 4. Having demonstrated that the numbers 1 through 120 could be written as the sum of four perfect squares, Bachet conjectured that it was possible to write *every* positive integer in this fashion. (For example, $120 = 64 + 36 + 16 + 4 = 8^2 + 6^2 + 4^2 + 2^2$. Note that the statement really means "four or fewer" squares, as you could substitute 10^2 for $8^2 + 6^2$ in the equation above.)

Pierre de Fermat (1601–1665) claimed to have a proof of Bachet's conjecture, but, in typical Fermat fashion, he never revealed it. The conjecture remained unresolved until 1770, when Joseph-Louis Lagrange published a proof.

▼

THE Four-Square Theorem falls in the category of existence theorems— although it proves that any positive integer can be written as the sum of four squares, it does not show how to get them. It takes a little poking and prodding to find out, for example, that $1,718 = 49 + 144 + 625 + 900 = 7^2 + 12^2 + 25^2 + 30^2$. And the representation doesn't have to be unique: $1,718$ also equals $1,600 + 81 + 36 + 1 = 40^2 + 9^2 + 6^2 + 1^2$. Nor do the squares have to be distinct—for example, the only way to get to 15 is via $9 + 4 + 1 + 1$. As it happens, the number 15 is not achievable using three or fewer squares, and in fact it is possible to identify those integers requiring the full four squares.

IN 1852, a young man named Francis Guthrie noticed that a map of the counties of England could be filled in using only four colors—while preserving the essential feature that no two adjacent counties (presumably, of the 39 historic counties, shown here in gray) be given the same color. Guthrie asked his brother Frederick if it was true that *any* map can be colored in this fashion. Frederick Guthrie then communicated the conjecture to his professor Augustus de Morgan (now known for the de Morgan rules of symbolic logic), and the battle was joined.

Like Fermat's Last Theorem, the Four-Color Theorem (or Four-Color Map Theorem) was called a theorem rather than a conjecture for many years. After a laborious process, and quite a few false starts, the theorem was finally proven—in 1977.

GERMAN mathematician Hermann Minkowski (1864–1909) is perhaps best known for introducing the technique of using geometry to prove results in number theory, and he is associated with the number 4 in two completely different respects. One is that the four-dimensional concept sometimes referred to as "space-time" (ordinary Euclidean 3-space—the kind we live in—together with a time component), is officially called Minkowski space. The second can be explained by use of the following diagram:

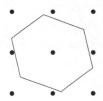

Assuming that the distance between any two adjacent dots (row or column) is 1, what can you say about the area of the hexagon? Well, it certainly appears that the area is less than the total area bounded by the dots, but Minkowski came up with a more tightly defined result: Any symmetric, convex region of area 4 must contain more than one lattice point. This result generalizes to n dimensions, and in the case of $n = 4$ it is possible to construct a special four-dimensional ball that leads to a quick proof of the Lagrange's Four-Square Theorem!

▼

I don't mean to close this discussion on a down note, but honesty compels me to acknowledge that some societies don't like the number 4 very much. In Mandarin, Cantonese, and Japanese, the words for "four" and "death" are apparently pronounced nearly identically, and the result is a cultural phobia for the ages.

5 [prime]

THERE are precisely five Platonic Solids—the name for a three-dimensional shape whose faces are all identical polygons. The most familiar Platonic Solid is the cube, whose faces are of course squares. The cube is joined by the tetrahedron, octahedron, and icosahedron (composed of equilateral triangles) and the dodecahedron (regular pentagons).

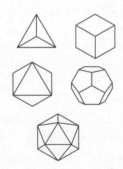

▼

EUCLID noted in his Elements (circa 300 BC) that these five solids were the only ones possible, the proof apparently credited to Theaetetus (see **2**). But at least some of the solids themselves were known quite a bit before that, judging by the photograph below, which shows carved Scottish stones believed to date to around 2000 BC.

THE world's largest office building is in the shape of a pentagon and goes by that name. Not only does the Pentagon in Arlington, Virginia, have five sides on the outside, it consists of five concentric pentagons with an interior courtyard that measures five acres. The pentagonal shape was originally deemed essential because of the contours of the original site of the building, but the shape wasn't changed even when Franklin D. Roosevelt opted for an alternative site. Roosevelt apparently took great interest in its architecture, praising the pentagonal shape because it had never been done before.

According to the official Pentagon website, the building itself covers 29 acres, bringing up an interesting coincidence that may not be coincidental at all. It starts with the observation that if you start with a regular pentagon, you can always create another, smaller pentagon by connecting vertices like so:

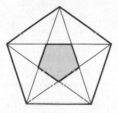

The length of a side of the inner pentagon turns out to be $\frac{(\sqrt{5}-1)^2}{4}$, or .038 times the length of a side of the outer pentagon, meaning that the inner area is $(0.38)^2 = 0.145$ times the area of the outer pentagon. Now, the five-acre courtyard inside the real, capital-*p* Pentagon is different in that it has the same orientation as the building itself, rather than the inverted orientation in the diagram. But the real-life relative area is $\frac{5}{(5+29)} = \frac{5}{34} = 0.147$, meaning that the real-life nested pentagons follow essentially the same proportions as the diagram we just drew.

IN yoga, the human body is often viewed as a pentagram, with the extremities defined by the head, two arms, and two legs. Positions such as the triangle pose involve five lines of energy. The number 5 also plays a role in less uplifting human processes, notably the Five Stages of Grief defined by nineteenth-century Swiss psychiatrist Elizabeth Kubler-Ross: (1) Denial and Isolation, (2) Anger, (3) Bargaining, (4) Depression, and (5) Acceptance.

IF you remember the quadratic formula from either your childhood or from the discussion in **2**, you know that an equation of the form $ax^2 + bx + c = 0$ has an explicit solution (or two), namely $x = \frac{(b \pm \sqrt{b^2 - 4ac})}{2a}$. The point is that the answer, x, is determined by the coefficients a, b, and c, as you'd expect. And there's a formula using radicals (i.e., square roots) for general cubic equations as well, though it's considerably more complicated than the quadratic formula. There's even a formula for the general quartic (fourth degree) equation, even though it takes that complexity several steps further. But there is no such formula for equations involving fifth powers. In other words, there is no way of writing down the solution(s) to the equation $ax^5 + bx^4 + cx^3 + dx^2 + ex + f = 0$ as a function of the numbers a, b, c, d, e, and f, using radicals. There's not even a hideously complicated way of approaching the problem. You just can't do it at all.

THE unsolvability of the general quintic equation is perhaps the most prominent conclusion that springs out of Galois theory, named for the Frenchman Evariste Galois (1811–1832). A quick look at Galois' entry and exit dates is a reminder of his precocity and of what might have been had he not taken up dueling at such a tender age. His final work wasn't published in full until 1846, fourteen years after his death at age 20. (The very first proof of the unsolvability of the quintic equation using radicals was accomplished in 1823 by the Norwegian mathematician Nils Henrik Abel. Abel was cursed in a different way, dying of tuberculosis at age 26.)

5 also serves as a limit for a completely different situation:

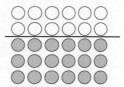

In the diagram above, checkers are placed below a line, with empty spaces above the line. Suppose you can advance the checkers above the line by leapfrogging over other checkers, as in the actual game of the same name. You can easily see that only two checkers and a single move are required to advance a man to the first rank above the line, as follows:

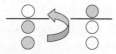

And you can advance a man to the second row if you had four originally, in the following three-move sequence:

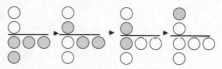

A little experimentation reveals that 8 men are sufficient to reach the third rank. If you've noticed the pattern so far (2^n men are required to reach the nth rank for $n = 1, 2, 3$) you might expect that 16 men will suffice for level 4, but in fact you need 20 men. The real surprise comes at level 5. No matter how many checkers you start with, you can *never* reach the fifth rank. This result is credited to John Conway, then of Cambridge University and now of Princeton.

SPEAKING of impossibility, it is appropriate to introduce the magic pentagram, which is basically a star. A standard five-pointed star has 10 vertices, each covered by a letter in the diagram. It would be aesthetically pleasing if you could replace each letter with a different number from 1 to 10 so that the sum of the numbers along each of the five lines was the same. But, as Charles Trigg first demonstrated in 1960, it just can't be done.

▼

WHEN you hear that the five Olympic rings represent the five regions of the world, it's easy to nod in agreement, but there are seven continents, not five, and only one of those (Antarctica) isn't represented at the Olympic Games. But the design is the brainchild of Pierre de Coubertin, founder of the modern Olympics, and it is apparently based on an old Greek artifact of similar shape.

The five regions of the world are Asia, Africa, Europe, the Americas, and Oceania. Although the colors of the rings—blue, black, yellow, green, and red—do not correspond to these regions in a one-to-one sense, each of these five colors is represented in every national flag in the world. Oceania doesn't have an Olympic team, but I would be remiss if I didn't mention a couple of nature's most marvelous pentagons: the sand dollar to the right and the sea star, or starfish. Five-fold radial symmetry is a characteristic of all echinoderms, a marine class that includes the sea star.

▼

SPEAKING of stars, my friend Norton Starr of Amherst College reminded me of one final appearance of the number 5 as a limit that I wanted to pass on. Consider a circle of radius 1, known in the trade as the unit circle. Because the area of a circle is given by the formula $A = \pi r^2$, the unit circle has area π. In three dimensions, the volume of a sphere is $(\frac{4}{3})\pi r^3$, so the unit sphere has volume $\frac{4\pi}{3}$. Although it is impossible to draw a unit sphere for n dimensions with $n > 3$, the concept is easily written down: Just as the unit circle is the set $\{x: x^2 \leq 1\}$, the general form for the unit sphere in n dimensions is the set of n numbers $\{x_1, x_2, \ldots, x_n\}$ such that $x_1^2 + \ldots + x_n^2 \leq 1$.

Our intuition tells us that the unit sphere keeps on growing in size, just as the unit box does—if only FedEx carried ten-dimensional boxes, we'd never have any trouble finding a box big enough for all our packages. But the punch line is not only that the volume of the *n*-dimensional unit sphere does not keep growing as *n* gets bigger. The real surprise is that this volume attains its maximum when $n = 5$, and steadily declines thereafter. In fact, the volume approaches *zero* as *n* heads to infinity.

6 $\left[\; 2 \times 3 \qquad 1 + 2 + 3 \qquad 1 \times 2 \times 3 \;\right]$

6 is the third triangular number, as evidenced by the triangle formation of the Blue Angels flight team.

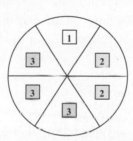

6 is also the first perfect number. A perfect number is a number that equals the sum of its proper divisors. In this case, $6 = 1 + 2 + 3$, or $\frac{1}{6} + \frac{2}{6} + \frac{3}{6} = 1$, as shown in the diagram. Perfect numbers are quite rare: 28 is the next one, followed by 496. To date, fewer than 50 such numbers have been identified.

AT sea, a depth of six feet is a fathom (derived from the Old English *faethm*, meaning outstretched arms). A fathom is likely associated with the expression "deep six," as in throw overboard. When on land, a depth of six feet is the traditional location of a casket, as in "six feet under."

DRAW six points on the plane (shown at right), and begin connecting each pair of points with either a black line or a gray line. By the time each pair of points is connected in this fashion, there will be at least one triangle that is either all gray or all black.

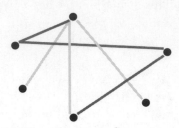

Note that the role of the number 6 is crucial. If you had only five points, no monochromatic triangle would be assured: For example, starting with a regular pentagon, you could color the perimeter gray and the interior diagonals black.

The proof of the above result doesn't require anything beyond simple arithmetic. Start by choosing any one of the six points. Consider the five line segments emanating from that point. Because there are only two colors, at least three of these lines must be the same color, say, gray. (As in the three gray lines from the top point in the diagram: The argument would be the same if they were black.) Now look at the points at the other end of these segments. If any two of these points were connected by a gray line, that line would complete a gray triangle. If not, they would all have to be black, creating a black triangle.

If this proof eluded you, don't worry. Back in 1953, an equivalent problem appeared in the William Lowell Putnam mathematics competition, an annual competition for math majors. The exact Putnam question was "Prove that in any group of six people there are either three mutual friends or three mutual strangers." Note that this problem is equivalent to the black and gray formulation above—for example, a black line connecting two points could represent friendship between two people.

SPEAKING of interpersonal friendships, the idea behind the now-popular phrase "six degrees of separation" is that any two people on earth can be connected through a chain of mutual friends that has no more than six links. The concept dates back to a 1929 short story "Chains" by Hungarian writer Frigyes Karinthy, and is not as preposterous as it sounds: Each new interpersonal link opens up a vast number of new connections, creating exponential growth.

Karinthy's concept was put to the test by social psychologist Stanley Milgram in a famous 1967 experiment, wherein random Midwesterners were each given a package to send to Cambridge, Massachusetts, guided only by the name, occupation, and rough location of the intended recipient. Without Google or switchboard.com to guide them, the only available strategy was to relay the package to whomever among their acquaintances they felt was most likely to get them close to their goal, and so on down the chain. Not all of the packages were delivered, but the median number of intermediaries for those that did make it was just five.

The "small world" concept came of age with John Guare's 1990 play *Six Degrees of Separation*, which was followed by a 1993 movie by the same name starring Stockard Channing, Will Smith, and Donald Sutherland. The Internet then created a few new wrinkles, including various attempts to update Milgram's experiment, and, most memorably, a game called Six Degrees of Kevin Bacon.

Bacon Number	Number of Actors/Actresses
0	1
1	1,879
2	158,022
3	447,500
4	109,360
5	8,178
6	863
7	93
8	13

Not only was Bacon's full name a perfect melodic fit, he was in so many well-known movies that most actors and actresses could be traced to him using three or fewer links, where the "0" link is Bacon himself, the "1" consists of the 1,879 performers who have acted in a movie with Bacon (as of 2004), and so on. The three Hollywood stars mentioned in the previous paragraph have "Bacon numbers" of 2, 2, and 1, respectively, Sutherland having appeared opposite Bacon in *JFK*. The average Bacon number within a database of 800,000 actors was a stunningly low 2.95.

Of course, once movie databases could be harnessed to prove how centered Kevin Bacon was within the Hollywood universe, those same databases revealed over a thousand performers who were even *more* centered. The most-centered actor of all—connected throughout the database with an average of just 2.67 links—turned out to be Academy Award–winner Rod Steiger.

There's actually an analogue to Bacon numbers within higher mathematics. Anyone who ever wrote a paper in collaboration with the prolific Hungarian mathematician Paul Erdos is said to have an Erdos number of 1, and so on down the chain. As with Bacon numbers, virtually all mathematicians have an Erdos number, and in virtually all cases that number is a single digit.

The ultimate achievement is a low Erdos-Bacon number, obtained by adding your Erdos and Bacon numbers together as if they were factored placements in a figure skating competition. (Or something like that.) Actress Danica McKellar (Winnie on *The Wonder Years*) also has a degree in mathematics from UCLA and an Erdos-Bacon number of just 4 + 2 = 6. But MIT professor Daniel Kleitman wrote a paper with Erdos and also appeared in *Good Will Hunting*, giving him an Erdos-Bacon number of just 3.

▼

THE great French mathematician Adrien-Marie Legendre (1752–1833), who developed among other things the *méthode des moindres carrés* (known in English-speaking statistics classes as the "least squares method" of linear regression/curve fitting), made a rare blunder by claiming that 6 could not be represented as the sum of the cubes of two rational numbers. (Obviously such a sum is impossible for two integers, and if irrational numbers were allowed you'd have trivial solutions such as 1 and the cube root of 5.) Legendre was no longer alive when Henry Dudeney, master British puzzlist of the late nineteenth and early twentieth century, found the surprisingly small counterexample of $6 = (\frac{17}{21})^3 + (\frac{37}{21})^3$.

SUPPOSE you had a map with a bunch of dots representing towns, and you drew lines connecting every dot with the closest dot to it. You might be surprised to hear that no town would have more than six lines emanating from it.

While the statement doesn't sound obvious, the proof is surprisingly easy and hinges on the simple fact that $\frac{360}{6} = 60$. Watch.

Suppose there were more than 6 lines coming from, say, Springfield. Then, since $\frac{360}{7} < 60$, two of those lines would have to meet at an angle less than 60 degrees, like so:

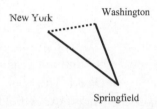

Now we have a triangle in which the angle at the Springfield vertex is less than 60 degrees. But if that's true, then at least one of the other angles is *greater* than 60 degrees, because the measures of the three angles of a triangle always sum to 180 degrees. In particular, the distance between New York and Washington would have to be less than the distance between either New York and Springfield or Washington and Springfield. (In the diagram, it's both.) But if New York and Washington are the closest together of the three, a line should join *them*, and the other lines shouldn't be there in the first place. The proof is now complete.

RELATED to the above discussion is the unique role played in the hexagon in two-dimensional tiling. Hexagonal tiling is actually quite familiar to us even if we've never heard the term, because it's the structure of chicken wire, honeycombs, and old-fashioned bathroom tiles.

BUT no regular polygon with more than six sides can possibly tile the plane. (Among regular polygons with *fewer* than six sides, equilateral triangles and squares easily tile the plane, but regular pentagons do not.) The key to tiling is found in the interior angles of the polygon. The interior angles of the equilateral triangle, square, and hexagon are 60, 90, and 120, respectively, and these numbers divide evenly into 360. In the diagram above, for example, three hexagons meet at each vertex, nicely filling the space and reminding us that 3 × 120 = 360. However, anything over six sides creates an angle measure of greater than 120 degrees, and the only proper divisor of 360 in this territory is 180, which is the angle measure of a line, not a shape. (The only other case is the pentagon, whose interior angles measure 108 degrees; again, it's impossible to shape a set of such angles around a vertex, so you're out of luck.)

IF you take two equilateral triangles and put them together just so, you get a hexagram known as the Star of David. The symbol of Judaism, the Star of David has been in use since the Middle Ages and has appeared on the flag of Israel since its founding in 1948.

IN mathematics, a Star of David Theorem refers to certain relationships within Pascal's Triangle (see **3**). Perhaps the simplest theorem to describe starts by surrounding one of the numbers in Pascal's Triangle with a tilted Star of David, as in the diagram.

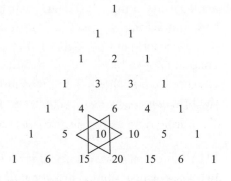

Each of the two triangles points to three different numbers. One of the triangles points to the numbers 5, 6, and 20, which have a product of 600. The other triangle points to the numbers 4, 10, and 15, whose product is also 600.

Yes, of course this always works. That's why it's a theorem. The results are even more remarkable-looking when you go farther down the triangle.

▼

EVERYONE knows that snowflakes are six-sided, but the first person to ruminate on the subject was Johannes Kepler. Legend has it that a snowflake fell on Kepler's overcoat as he was crossing a bridge over the Vltava river in 1611, and he decided to write a treatise on snowflakes as a gift to his patron. Kepler's view was that the abundance of sixes in nature, from flower petals to honeycombs to snowflakes, arose from deep-seated structural principles—which he dubbed *facultas formatrix*—associated with the efficiency of hexagonal packing. Apparently Kepler was something of a punster, as he wrote his treatise in Latin, in which the word for snowflake is *nix*, which in turn meant "nothing" in his native Lower German. The fact that his treatise cost nothing appealed to Kepler, chronically short of funds because of the stinginess of Emperor Rudolph II, to whom Kepler served as the imperial astronomer.

7 [prime]

THE number 7 isn't quite as easy to cope with as any number we have seen thus far, because the arithmetic of 7 isn't characterized by obvious patterns. That property actually makes 7 useful to neurologists: One of the basic tests for the onset of dementia is to have the patient successively subtract 7 from 100. The sequence runs 100, 93, 86, 79, 72, etc., and clearly provides more of a hurdle than using, say, five instead.

▼

ONE nice pattern produced by 7 that 5 cannot match is the fraction for $\frac{1}{7}$: 0.142587.
Get a load of the multiplication table for the repeating portion of the number:

142857 ×	
2	285714
3	428571
4	571428
5	714285
6	857142
7	999999

The number 142857 is called a cyclic number, because you obtain its successive multiples by beginning with one of its other digits and wrapping around, maintaining the same order. The key to the magic is that the repeating portion of the decimal expansion of $\frac{1}{7}$ has six digits. The next fraction to have this property is $\frac{1}{17}$, which equals 0.0588235294117647. (Note that the repeating portion now has 16 digits. In general, if the fraction $\frac{1}{p}$ has a decimal expansion with a repeating sequence of length $(p-1)$, that sequence forms a cyclic number.)

▼

THE Seven Bridges of Konigsberg is a topological problem (indeed, perhaps the *first* topological problem) studied by Leonhard Euler. The legend has it that Euler himself traipsed across the bridges of this old Prussian town (now a part of Russia and renamed Kaliningrad in 1946). The bridges connected various parts of the town separated by rivers, and the question of the day was whether it would be possible to walk across all seven bridges without retracing his steps.

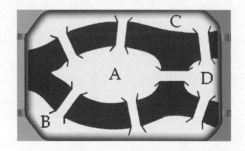

Euler eventually realized that such a crossing was impossible. The analysis of the problem is made easier by a diagram focusing on the bridge structure. The impossibility of an Eulerian path that encompasses all the bridges is found in the node structure of the bridges; namely, the number of entrances and exits from the points A, B, C, and D.

It turns out that an Eulerian path is only possible if there are exactly two or zero nodes of odd degree. In Konigsberg, however, all of the nodes had odd degree—five for region A and three for the other three. As mathematically important as Euler's walks through Konigsberg were, the daily walks of Konigsberg resident Immanuel Kant were memorable in a different way. Legend has it that Kant was so punctual in his appearances around town that locals learned to set their watches by him.

Seven Famous Sevens						
Continents	Days of the Week	Hills of Rome	Wonders of the Ancient World	Wonders of the Modern World	Colors of the Spectrum	Deadly Sins
Asia	Sunday	Aventine	The Great Pyramid of Giza	Empire State Building	Red	Lust
Europe	Monday	Caelian	The Hanging Gardens of Babylon	Itaipu Dam	Yellow	Gluttony
Africa	Tuesday	Capitoline	The Temple of Artemis at Ephesus	CN Tower	Blue	Greed
Australia	Wednesday	Esquiline	The Statue of Zeus at Olympia	Panama Canal	Green	Sloth
North America	Thursday	Palatine	The Mausoleum at Halicarnassus	Channel Tunnel	Indigo	Wrath
South America	Friday	Quirinal	The Colossus of Rhodes	Delta Works	Violet	Envy
Antarctica	Saturday	Viminal	The Pharos of Alexandria	Golden Gate Bridge	Orange	Pride

THE dots on opposite sides of an ordinary die must sum to 7. In cards, it is a commonly held belief that a series of 7 shuffles will result in a deck being fully randomized.

THE frieze design is characterized by repeated copies of a single pattern in one direction. Friezes are often seen in wallpaper borders or perhaps as part of an architectural flourish in an old building. Although we have all seen plenty of examples of friezes, the word itself is not universally known and more obscure still is the fact that there are fundamentally only seven different types of symmetries that a frieze can possess. Princeton's John Conway introduced footprints as a means of distinguishing among these seven "one-directional" symmetries, a practice we will follow in the list below:

1. **THE HOP** (a simple translation symmetry)

2. **THE STEP** (translation and glide reflection symmetries)

3. **THE SIDLE** (translation and vertical reflection symmetries)

4. **THE SPINNING HOP** (translation and half-turn rotational symmetries)

5. **THE SPINNING SIDLE** (translation, glide reflection, and half-turn rotation)

6. THE JUMP (translation and horizontal reflection)

7. THE SPINNING JUMP (translation, horizontal and vertical reflection, and rotation)

THE two diagrams to the right illustrate the Seven Circles Theorem. If you start with a circle and draw six circles tangent to that circle, six points of tangency are created. Remarkably, those six

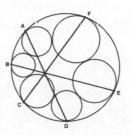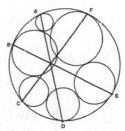

points can be separated into three pairs of two points such that if you draw a line connecting each of those three pairs, the three lines intersect in a single point. That's true regardless of what configuration happens to be involved.

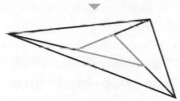

START with a triangle (above), and extend each of the three legs out until it is twice its original length. The result is a new triangle whose area is precisely seven times the area of the triangle you started with. This is because if you connect any vertex of a triangle at the midpoint of the opposite side, you have created two triangles of equal area. Drawing a few extra lines into the diagram enables us to create seven triangles, all of which have equal area, which is another way of saying that the center triangle is one-seventh the size of the large triangle.

▼

FINALLY, the diagram to the right is called a "seven segment display." Its name is not well-known but it is a standard means of displaying the ten digits 0 to 9 as well as the letters A, b, C, d, E, and F. Note that b and d are in lowercase, and that's because their capital forms can't be distinguished from 8 and 0 respectively. The seven segment display is often slanted to improve readability, but in this form it is in the shape of an 8, our very next subject.

8 $\left[2^3 \right]$

THE number 8 owes many of its appearances to the fact that it is a power of two. For example, cutting a block in half along each of its three axes will produce eight smaller blocks. But 8 also appears within an entirely different formulation.

Consider the humble though famously fertile rabbit. Actually, start with a pair of rabbits, one male and one female. The reproductive ground rules are (1) any pair of rabbits becomes fertile after one month, and (2) the pair delivers another pair one month later, and every month after that. As long as the rabbits don't die, so to speak, here's how the generations proceed:

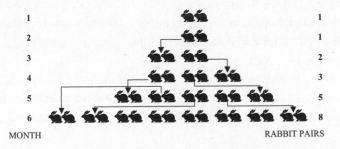

MONTH   RABBIT PAIRS

The numbers in the rightmost column form the famous Fibonacci sequence, named after Leonardo de Pisa (circa 1175–1250), aka Fibonacci, who was apparently the first person in the Western world to study the sequence and the one who is credited with originating the rabbit metaphor. The Fibonacci sequence turns out to have many other interpretations (the number of ways of climbing an n-step staircase using either one or two steps at a time, the number of coverings of a $2 \times n$ checkerboard by 2×1 dominoes, and so forth), but we don't need those or even the rabbit labyrinth to keep the sequence going, because we can see at a glance that any member of the Fibonacci sequence is the sum of the previous two members of the sequence. (In math talk, this is an example of a recursively defined sequence.)

If you are troubled by the inbreeding of the rabbits required to sustain the Fibonacci sequence, consider this paradox of sorts. When we scan our family tree past our two parents, four grandparents, and even our eight great-grandparents, we expect 16 in the generation before all of those, 32 in the one before that one, and so on. But you can't very well keep doubling your ancestors, can you? There are *more* people alive today than centuries ago, not fewer. In particular, if you figure that Fibonacci himself lived 30 generations ago, there is simply no way that we had 2^{30} (more than one billion) ancestors from that time, because there weren't that many humans on the planet. What's going on? The inescapable conclusion is that there were plenty of duplications among these 2^{30} forebears, so there was inbreeding, all right, not necessarily the scandalous kind of our rabbit model, but a bit closer to home.

You will have noticed by now that 8 is both a Fibonacci number and a perfect cube. Eight is actually the last cube in the Fibonacci sequence, the only other one being 1. What's more, 8 is the largest Fibonacci number that has a prime neighbor—namely, 7. (We will see more Fibonacci numbers before this book is through, and you can check out all of their neighbors to make sure a prime number doesn't show up.)

▼

IN the United States, the stop sign has officially been an octagon since the 1920s, the idea being that motorists could recognize its distinctive shape even from the reverse side. Most of the English-speaking world uses the same construction.

▼

THE geometry of the number 8 goes well beyond octagons. For example, a square can be subdivided into 8 acute triangles (all angles less than 90 degrees), as in the diagram to the right. Seven triangles do not suffice.

▼

SPEAKING of squares and the number 8, if you put an asymmetric diagram inside a square, you can produce precisely eight different images via rotations and reflections, as below. (The top-left figure is rotated by 90, 180, and 270 degrees to produce the three images to its left, and the top row is reflected from top to bottom to produce the bottom row.) In math-speak, the diagram is being acted upon by the dihedral group of order eight. In general, the dihedral group associated with a regular polygon of n sides has order $2n$.

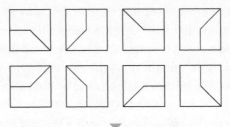

▼

IT is commonly believed that seven shuffles are the right amount to produce a random order of 52 cards, but eight perfect riffle shuffles will return a 52-card deck to its original order.

▼

SPIDERS have eight legs and, usually, eight eyes. But the eight-leg part is what distinguishes them from insects.

▼

IF an octagon has eight sides, an octopus has eight legs, then surely October is the eighth month of the year. No, wait a minute, that's not right. October is the tenth month. What happened? The answer is that October was indeed the eighth month of the Roman calendar, but that was before January and February were officially christened to occupy what had been a monthless winter period.

▼

IN the sport of rowing, an "eight" typically refers to an eight-person crew (coxswain not included), but the number 8 figures into a wide range of other sports:

For example, running tracks and swimming pools typically have eight lanes. (In both cases, not all lanes are created equal; some positions are known to be more advantageous than others.)

The middle lanes were familiar territory to the USA's Michael Phelps as he swept to a record eight gold medals in the 2008 Summer Olympic Games. Those games, of course, began on 8/8/08, in keeping with the long-standing reverence for the number 8 in Chinese numerology.

The "figure eight" is commonly associated with figure skating, an eight being one of many different patterns that competitive skaters once had to trace with their blades. Fittingly, the last time that so-called compulsory figures were seen in the Winter Olympics was in '88. The very next year saw a rather quirky appearance of the number 8. At the conclusion of the 1989 Tour de France, after American Greg Lemond came from behind on the final day's time trial to win the Tour by a grand total of eight seconds, runner-up Laurent Fignon could be seen lying on the ground in exhaustion and disbelief. The wheels of his bicycle, lying beside him, formed a perfect 8.

▼

THE idea of a wheel with eight spokes goes back to the Dharmacakra, one of the eight auspicious symbols of Buddhism. And the idea of using two circles to represent the number 8 is best known in golf, giving rise to the colorful term *snowman* to represent an 8 on a particular hole—no better than a triple bogey.

9 [3^2]

THE equation $9 = 3^2$ is a nice follow-up to $8 = 2^3$ and it's one of a kind. Never again will you find two consecutive numbers that are both perfect powers, much less perfect powers having the beautiful symmetry of 2^3 and 3^2.

▼

IN American sports, a "nine" is a baseball team, and the number is imbued throughout the game. When using an official scorecard to follow a game, for example, each defensive player is given a number, from 1 (the pitcher) to 9 (the right fielder). In any particular half inning, there are three outs, and given that three strikes make an out, it is possible for a half inning to consist of nine strikes and nothing else (a feat that has happened more than 40 times in the history of Major League Baseball). And if a team doesn't show up for a game, that game is forfeited, and in the official score the winner is given a number of runs equal to the number of innings in a game, also nine. That's right. A no-show in a major-league baseball game goes into the record books with the score 9–0.

▼

ONE of the most familiar depictions of 9 as three squared comes from a different game, tic-tac-toe, also known as noughts and crosses.

Tic-tac-toe is of course a very simple game. Unless you've never seen it before, you'll have no difficulty in achieving a draw no matter whom you're playing against. Or even what. Ginger the tic-tac-toe playing chicken was an instant hit when she reached Las Vegas in 2002, having apparently played for nine months at Atlantic City's Tropicana Hotel and only lost five times.

▼

A different expression of 9 as three 3's comes from "the nine Worthies," a group of historical/legendary figures designated in the Middle Ages. The Worthies were as follows:

Pagan	Jewish	Christian
Hector	Joshua	King Arthur
Alexander the Great	David	Charlemagne
Julius Caesar	Judas Maccabeus	Godfrey of Bouillon

Together, the nine Worthies supposedly represent all facets of a perfect warrior.

▼

SOME of the most important properties of the number 9 stem from its being just 1 less than 10, the base of the number system. Perhaps the best known of these properties is the test to determine whether or not a number is divisible by 9: Just add the digits of the number, and if the resulting sum is divisible by 9, so is the original number.

For example, if you add the digits of the number 176,328 you get $1 + 7 + 6 + 3 + 2 + 8 = 27$. Because 27 is divisible by 9 (you could repeat the process to get $2 + 7 = 9$, which certainly is divisible by 9), so is 176,328.

▼

THE technique of checking your arithmetic by "casting out the nines" is based on a similar idea. Suppose you've just performed the following addition:

$$\begin{array}{r} 1428 \\ + 5837 \\ \hline 7255 \end{array}$$

If you add the digits of 1428 and "cast out the nines," you're left with $1 + 4 + 2 + 8 - 9 = 6$. (In math talk, 6 is the "digital root" of 1428.) If you do the same with 5837 you get 5. And if you add 6 and 5 and cast out one final nine, you get 2. If you've done the addition properly, you should get the same result if you add the digits in your sum and cast out the nines. Unfortunately, that doesn't work here, because $7 + 2 + 5 + 5 - 9 - 9 = 1$. Something has gone wrong. Upon double-checking the original problem, we see that we made an error in the tens column, when the one wasn't carried from the sum $8 + 7 = 15$. The answer should have been 7265.

Note that "casting out the nines" cannot guarantee that your original arithmetic was correct. If you were wrong originally, however, the technique gives you a shortcut in discovering that an error was made.

▼

ALL this is rather tedious stuff, but consider that the number 9 and its digital roots are core building blocks for the likes of Arthur Benjamin—professor of mathematics at Harvey Mudd College during the day, mathemagician by night. Benjamin routinely fools his audiences with math "tricks" that involve nothing more than the basic properties of 9 and so-called digital roots in the spirit of casting out the nines. Example: Pick a four-digit number. Scramble its digits to form a new number, and subtract the smaller one from the larger one. From this new number, which we'll call N, take away a nonzero digit. Upon hearing the remaining digits, it's child's play to figure out the missing digit, because the digital root of N must be 9.

▼

LONG after teachers stopped advising students to cast out the nines, they were still telling students that there were nine planets: Mercury, Venus, Earth, Mars, Jupiter, Saturn, Uranus, Neptune, and Pluto. Unfortunately, as everyone knows by now, Pluto has been demoted to "dwarf planet" status, where it joined the likes of Ceres and Eris. Why the International Astronomical Union waited until August 2006 to officially define a planet in a way that excluded Pluto, I have no idea, but I suppose it was their way of casting out the ninth one.

▼

ONE group of nine that is not destined to change is the Nine Muses, together with their specialties:

Calliope: Epic poetry
Clio: History
Erato: Love
Euterpe: Music/lyric poetry
Melpomene: Tragedy

Polyhymnia: Sacred songs
Terpischore: Dance
Thalia: Comedy/bucolic poetry
Urania: Astronomy

▼

ANOTHER more recent group of nine is the US Supreme Court, set at its current number in 1869. The group was referred to as the "nine old men" by Franklin Roosevelt, who went so far as to propose in 1937 that the president should be able to appoint an additional justice for every one over 70½ years of age. The reason given for the change was to reduce the workload, but Roosevelt's thinly disguised aim was to pack the court with justices who were less likely to declare his various New Deal proposals unconstitutional. Despite making it the subject of the first of his nine fireside chats, FDR's court-packing plan went absolutely nowhere, but because his presidency was so long, he was able to appoint a total of eight justices to the court, the biggest number of appointments since George Washington.

▼

WE'LL close with a magical appearance of 9 that may be less familiar. Start by drawing an acute triangle. (Others will do, but an acute triangle works out the best.)

Figure 1

Figure 2

Figure 3

Figure 4

1. Mark the midpoints of each side (3 points). See Figure 1.
2. From each vertex, drop an altitude (a line that is perpendicular to the side opposite the vertex). Mark the points where the altitudes intersect the opposite side. See Figure 2.
3. Notice that the altitudes intersect at a common point. Mark the midpoint between each vertex and this common point. You have created a total of nine points. See Figure 3.

4. No matter what shape triangle you started with, these nine points all lie on a perfect circle! See Figure 4.

10 $[\,2 \times 5\,]$

WRITING about 10 is a little bit like writing about 1. It's everywhere, and what makes the number 10 special is kind of the same thing as what makes it less than special. We take it for granted.

The number 10 is best known as the base of our number system. The arithmetic of 10 is especially easy, in that $10^2 = 100$, $10^3 = 1000$, and in general 10^n equals a one with n zeroes after it. In particular, when we say that an estimate is off by an order of magnitude, technically that means it's off by a factor of 10, though real-life usage isn't always that exact. Although there are many other bases/number systems in life and in this book, 10 could be considered the base of any number system, in the sense that the number n, when written in base n, is always 10.

▼

THE sequence EOEREXNTEN ends with TEN. What does the sequence represent? (See Answers.)

▼

10 is a triangular number, as anyone who has ever bowled can tell you. But 10 is really, really triangular. Not only is it the sum of the first four integers, it is also $1 + 3 + 6$ – the sum of the first three triangular numbers. This last property makes 10 a tetrahedral number, meaning that you can build a tetrahedron by stacking 10 spheres—bowling balls, say—on three levels.

▼

THE figure below is not the familiar triangular arrangement of the ten bowling pins. It is the Tetraktys of the Pythagoreans, those spiritual followers of Pythagoras who were especially taken with the fact that $10 = 1 + 2 + 3 + 4$.

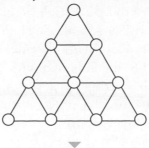

THE list below is another link between Pythagoras and the number 10:

The 10 Principles of Pythagoras	
(Also known as the table of the Opposites)	
limit	unlimited
odd	even
one	plurality
right	left
male	female
at rest	moving
straight	crooked
light	darkness
good	bad
square	oblong

So we see that the Top 10 list is an ancient concept. It seems, however, that the Pythagoreans weren't above a little shading and fudging to make their lists the right length: Said Aristotle, "They (the Pythagoreans) say that the bodies which move through the heavens are ten, but as the visible bodies are only nine, to meet this they invent a tenth—the 'counter-earth'" (*Metaphysics* 906a 10–12).

SPEAKING of triangles, the dissection to the right is the handiwork of William Gosper, a mathematician and programmer who is sometimes credited as being one of the original computer hackers. In this particular creation, a square is subdivided into 10 isosceles triangles (two sides equal), the minimum possible. (Compare **8**, the minimum number of acute triangles, not necessarily isosceles, into which a square can be subdivided.)

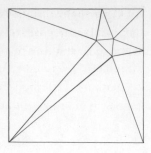

EVERYONE knows about the Ten Commandments, but less well-known is that there were only 10 original Rorschach inkblots, each with an unbounded number of possible interpretations.

THE prefix *dec-* means "10" in a whole range of contexts. A decagon has 10 sides, a decimal point refers to a place in our base 10 numbering system, a decathlon has 10 events, and so on. Less well-known is the word *decimate*, which is now used interchangeably with such words as *destroy* and *annihilate*, even though a quantity that has been decimated has, technically speaking, only been diminished by one-tenth of its original size. Apparently the word dates back to ancient Rome, when the punishment for mutiny was to kill one mutinous soldier out of ten. The fact that *decimate* has morphed into "utter destruction" suggests that the original policy of decimation had a powerful deterrent effect.

THE equation $10! = 6! \times 7!$ is a special one. While it is not too hard to construct an infinite family of factorials that are products of other factorials (in general, if $A = B!$, then $A! = (A-1)! \times B!$), you won't find any others where two *consecutive* factorials are being multiplied.

WE'LL close the discussion with an even more remarkable equation, one that represents any positive integer n using only the 10 digits 0 through 9, in order! (The number of square roots in the expression below equals n.) The formula is attributed to Verner Hoggatt Jr., a mathematician best known for his work on the Fibonacci series. Readers familiar with the definition of a logarithm are invited to ponder why the formula works.

$$\log_{(0+1+2+3+4)/5}\left(\log_{\sqrt{\cdots\sqrt{(-6+7+8)}}} 9\right) = n$$

(See Answers.)

11 [prime]

J. J. SYLVESTER, the British born founder of the *American Journal of Mathematics* and the onetime mentor of ace statistician turned nursing pioneer Florence Nightingale, proved in 1884 that the highest number that can't be created out of the numbers x and y equals $xy - x - y$. In particular, 11 is the highest number that cannot be attained as a score in rugby union using only drop goals (3) and converted tries (7), because $11 = 3 \times 7 - 3 - 7$. The same calculation applies to American football, using field goals and touchdowns with one-point conversions. That's only fair, as each team has 11 players on the field at any given time, the same as in soccer and cricket.

▼

ONCE you've seen Sylvester's formula, you should be able to absorb the similar-looking formula $\frac{1}{2}(x-1)(y-1)$ as the total number of scores that cannot be attained using x and y. Sure enough, whether you're talking about rugby or football or just numbers, you can't get to 1, 2, 4, 5, 8, or 11 using combinations of 3 and 7, and that's a total of 6 numbers, just as the formula $\frac{1}{2}(2)(6) = 6$ predicted.

▼

AN 11-sided figure is called a hendecagon. Surely the most famous hendecagon of all time is the one inscribed within the Susan B. Anthony dollar, first issued from 1979 through 1981 and then again in 1999. Originally, the coin itself was supposed to have been a regular hendecagon, but vending machine manufacturers never got around to ac-commodating any shape other than round. Unfortunately, without an 11-sided exterior to set it apart, the Susan B. Anthony dollar was frequently mistaken for a quarter.

THE number 11 provides some curiosities in the areas of multiplication and division. To begin with, the two-digit multiples of 11 are easy to spot, because they consist of repeated digits: 11, 22, 33, all the way up to 99. Three-digit multiples of 11 don't stand out quite as much, but they are surprisingly easy to generate. For example, start with 3 and 4. Add them to get 7. Put the seven in the middle of the 3 and 4 to form the three-digit number 374. That number is divisible by 11. In particular, 374 = 34 × 11. The appearance of the 7 is even more logical when you write out the multiplication in old-fashioned grammar school form and examine the middle column:

$$
\begin{array}{r}
34 \\
\times\ \underline{11} \\
34 \\
+\ \underline{340} \\
374
\end{array}
$$

The above procedure doesn't generate all three-digit multiples of 11, because it depends on the two chosen integers (3 and 4 above) not adding up to more than 9. If you started with 5 and 8 instead of 3 and 4 and followed the above rules literally, you'd get 5138, at which point you'd have to add the 5 and 1 to get the actual product of 638.

THAT same principle of carrying extends to the following triangle:

The five rows of the triangle happen to coincide with the first five powers of 11: $11^0 = 1$, $11^1 = 11$, $11^2 = 121$, $11^3 = 1331$, $11^4 = 14641$. But in fact this is the beginning of the famous Pascal's Triangle, in which 1's appear on the diagonal and each inner number is the sum of the two numbers above it, to the left and to the right. The next row of Pascal's Triangle is 1 5 10 10 5 1, not $11^5 = 161051$. Note that 11^5 is the first power of 11 that is not a palindrome.

THE general rule for divisibility by 11 works like this: Add up the digits of the number located in an odd position, then add the remaining digits. If the difference between these two sums is a multiple of 11, including 0, then the original number is divisible by 11. For example, the number 42,658 is divisible by 11 because $(4 + 6 + 8) - (2 + 5) = 18 - 7 = 11$.

NOW for a completely different multiplicative property that borders on the unbelievable. Take any number and multiply its digits together. Whatever number results, multiply *its* digits together, and keep going. Eventually, you'll get to a single-digit number. Sometimes the process is swift. For example, if the original number has a 0 in it you get to 0 immediately, while if the number has a 5 and any even digit you'll get to zero in two steps. But sometimes things take a bit longer. For example, though not by chance, if you start with the number 277,777,788,888,899 you get the following chain:

Step	Number	Product of Digits
1	277,777,788,888,899	4,996,238,671,872
2	4,996,238,671,872	438,939,648

Step	Number	Product of Digits
3	438,939,648	4,478,976
4	4,478,976	338,688
5	338,688	27,648
6	27,648	2,688
7	2,688	768
8	768	336
9	336	54
10	54	20
11	20	0

If you're thinking that 11 steps is a lot, you're on the right track. Remarkably, *no number* is known to require more than 11 steps, and it's not as if people haven't been looking. In 2001, Phil Carmody confirmed that all numbers less than 10^{233} have a "multiplicative persistence," as it is called, of less than or equal to 11. Our chosen number of 277,777,788,888,899 is the *smallest* number with a multiplicative persistence of 11.

▼

THE number 11 pops up behind the scenes in some clock problems. For example, start at 12:00 noon, when the hands of a clock are pointing in the same direction. Sixty-five and $\frac{5}{11}$ minutes later—in other words, one-eleventh the time between 12:00 noon and midnight—the hour and minute hands will again be pointing the same way.

▼

AN "Ask Marilyn" column (*Parade*, May 6, 2007) asked readers to fill in the time that is missing from the following sequence: 1:38, 2:44, 3:49, 4:55, ___, 7:05, 8:11, 9:16, 10:22, 11:27, 12:33. While it is tempting to look for patterns within the times themselves, once you notice that there are 11 times altogether, you'll shift to another tack.

The puzzle, called to Marilyn vos Savant's attention by Jacob Miller of Mount Joy, Pennsylvania, is a variation on an old theme. While the times don't appear to have anything to do with one another, that's just a casualty of the changeover from analog to digital watches (the bigger casualty being the heightened difficulty of being able to tell the time by catching a glance of someone else's wristwatch). The 11 times of day in the puzzle would stand out on an analog watch as being those occasions when the minute hand and hour hand point in precisely opposite directions. The missing time, obviously, is 6:00, which also happens to be the only time that comes already rounded off.

Let's look at this situation another way. Starting with and including 6:00, how many times will the hour and minute hands be precisely 180 degrees apart before the next 6:00? This obviously happens every hour, except that 6:00 is the only occurrence between 5:00 and 7:00. And clearly the time between any two successive occurrences is the same, because the phenomenon is a function of the relative speeds of the two hands, which never changes. Therefore you get a 180-degree spread every $\frac{12}{11}$ hours, or every one hour, five minutes, and $\frac{5}{11}$ seconds. (Sorry, but you just can't get rid of the 11 in the denominator. In particular, the times in the "Ask Marilyn" version are necessarily rounded off.)

▼

THE number 11 shows up in two geometric counting exercises. The 11 figures here represent the ways in which the edges of six squares can be connected so that the resulting two-dimensional shape can be folded into a cube. Six-squared figures of this type are called hexominoes, of which there are 35 altogether, assuming that reflections and inversions are not considered distinct. The 11 special ones are called the nets of the cube.

Objects such as hexominoes get plenty of mathematical attention for their tiling

properties. It turns out that the 11 hexominoes above cannot tile a rectangle, but the number 11 has a different and quite wonderful association with planar tilings, also known as tessellations.

We have already seen that there are just three regular polygons (the hexagon, the square, and the equilateral triangle shown on the left below) that can tile the plane. But if you can mix *different* regular polygons, the number of tilings increases to 11. These 11, one of which appears in the middle below, are the so-called Archimedean tilings of the plane, though Archimedes himself didn't have a whole lot to do with them. They were apparently studied and classified by Johannes Kepler, who for the record lived some 1800 years after Archimedes' time. The Archimedean tiling found in the middle combines regular hexagons and equilateral triangles.

In what at first glance appears to be an extraordinary coincidence, there are also 11 fundamentally different tilings of the plane by identical, convex, symmetric polygons (there are a couple of other restrictions, but that's enough math jargon for a family publication). The right-most tiling, made up of identical irregular pentagons (recall that *regular* pentagons cannot tile the plane) belongs in that category:

There is a simple one-to-one correspondence between the two sets of tessellations. If you start with an Archimedean tiling, mark the center of each polygon and then connect each dot to its neighbor(s), you obtain a tiling called the "dual" of the one you started with. In the layout on this page, the tessellation on the far right is the dual of the Archimedean tiling to its left. The dual tilings are called laves, or at least that's what I first thought. In fact they are Laves tilings, named after Swiss crystallographer Fritz Laves. Oh, well. Note that the standard hexagonal tiling is the dual of the equilateral triangle tiling shown here (and vice versa), while the standard square tiling (not shown) has the distinction of being self-dual, if I may sneak in one more mathematical term.

Although these various tilings justify their study by their intrinsic beauty alone, the Laves tilings are also studied in the science of materials: crystals, metallic alloys, and the like. Nature may abhor a vacuum, but it loves symmetries, and many of the designs on the previous page arise in many natural (though sometimes microscopic) contexts.

12 $\left[\, 2^2 \times 3 \,\right]$

THE number 12 is a favorite in many religions, what with the 12 days of Christmas, Twelfth Night, the 12 Apostles, the 12 feasts of Eastern Orthodoxy, and the 12 Tribes of Israel. These 12 tribes are associated with the 12 sons of Jacob, so it bears mention that the Norse god Odin also had 12 sons.

▼

THE concept of a dozen was also alive and well in Greek mythology, as evidenced by the 12 principal gods of the Greek pantheon atop Mount Olympus. These gods won't be listed here because their total number actually exceeded 12, but 12 was apparently the limit at any given time. However, we *can* list the twelve labors forced upon Hercules by King Eurystheus of Tiryns. Most involved killing some ghastly creature or another. One notable exception was labor number three, as the Cerynian Hind was actually a delicate deer, loved by Artemis, which Hercules had to stalk for a year before gently carting it away.

The Twelve Labors of Hercules
One: Kill the Nemean Lion
Two: Kill the Lernean Hydra
Three: Capture the Cerynian Hind
Four: Capture the Erymanthian Boar
Five: Clean the Augean Stables
Six: Kill the Stymphalian Birds
Seven: Capture the Cretan Bull
Eight: Capture the Horses of Diomedes
Nine: Take the Girdle of the Amazon Queen Hippolyte

Ten: Capture the Cattle of Geryon
Eleven: Take the Golden Apples of the Hesperides
Twelve: Capture Cerberus

▼

ON the other hand, whereas today it is standard for a jury to consist of 12 people, Ancient Greece had no such limit. The trial of Socrates involved 501 jurors.

▼

ONE reason for the prominence of the number 12 is that it is evenly divisible by 2, 3, 4, and 6, and is therefore convenient for all sorts of applications, from eggs to donuts to numbers on a clock to months in a year to signs of the zodiac.

▼

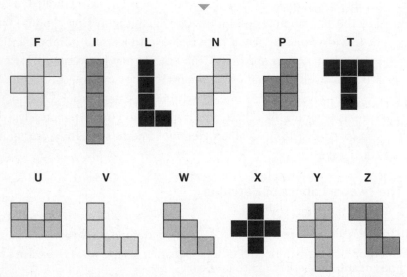

THERE are 12 different shapes that can be made out of five squares joined at the edges. Called pentominoes, these shapes are frequently labeled by the letters they most closely represent. Altogether, the 12 pentominoes account for 5 × 12 = 60 square units, and in fact it is possible to arrange the 12 pieces so as to make rectangles measuring 6 × 10, 5 × 12, 4 × 15, and

3 × 20. There is even a board-game variation in which players alternate placing pentominoes onto an 8 × 8 grid until someone (i.e., the loser of the game) cannot place a remaining pentomino without overlapping one that has already been played. Author and futurist Arthur C. Clarke was a big fan of pentominoes, and for *2001: A Space Odyssey* he worked in a scene that showed HAL the computer playing the 8 × 8 pentomino game. Unfortunately for fans of the game, the scene was cut in favor of a different 8 × 8 game: chess.

12 is also the "kissing number" in three dimensions. To see what this is all about, it might help to visit the two-dimensional case, in which the kissing number is defined as the number of circles of radius 1 that can simultaneously touch a given circle. That number is clearly 6, as suggested by the diagram at right.

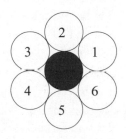

THE situation in three dimensions is necessarily more complex. When you put 12 spheres around a central sphere, there is substantial free space, and it is tempting to believe that there might be room for a thirteenth sphere.

Apparently the possibility of a thirteenth sphere arose in a conversation between Isaac Newton and Scottish astronomer David Gregory back in 1694. Although the precise nature of their disagreement has been lost to history (most accounts suggest that Newton was on the 12 side and Gregory on 13), there is no record of so much as a gentleman's wager on the subject and certainly no possibility of anyone collecting, as the question wasn't fully resolved until 1953.

SPEAKING of spheres, the traditional black pentagon/white hexagon Telstar soccer ball design was the official ball for the 1970 Mexico City World Cup and for several years thereafter. (1970 was the first year in which the

World Cup was televised, and the new ball was especially easy to follow on television.)

There are 12 pentagons and 20 hexagons on this ball. So why is this discussion being held under 12 rather than 20? Because the Telstar ball is actually a special case of a more general concept called a Buckyball, named for Buckminster Fuller of geodesic dome fame. (I say this rather sheepishly and ironically, as Buckminster Fuller was the speaker when I graduated from high school and I had never heard of him.) It turns out that many different sphere-like structures can be made from a combination of pentagons and hexagons, but whereas there is no limit on the number of hexagons, the number of pentagons must always be 12. (This follows from Euler's Theorem and the reader is invited to take a stab at a proof. See Answers.) In particular, it is possible for the number of hexagons to equal zero, in which case you're left with the regular dodecahedron, one of the five Platonic solids (see **5**). There is also a dodecahedron whose faces are all rhombi. Rhombic dodecahedra, as they are called, can, like cubes, be fitted together to fill three-dimensional space, in the same sense that rhombi—or squares, but not pentagons—can tile the plane. But a desk calendar can be made out of either form of dodecahedron, with each face getting a different month.

▼

AN alexandrine, found in French literature from Racine to Baudelaire, is a line of 12 syllables. Alexander Pope, alas, was not taken with this virtually eponymous form, as he writes:

A needless alexandrine ends the song
that like a wounded snake, drags its slow length along

▼

THE Necker cube is a 12-lined optical illusion that first appeared in 1832. Because the cube has no dotted lines, the figure produces (at least) two different interpretations.

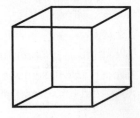

13 [prime]

THE number 13's biggest claim to fame is that it is considered exceedingly unlucky. There is no single reason why 13 is treated the way it is, but most explanations begin with the fact that it compares poorly with 12, its immediate predecessor. Twelve is nicely divisible by 2, 3, and 4, and shows up in months of the year, signs of the zodiac, and a whole range of other designations, as we have just seen. Thirteen, on the other hand, is prime and therefore quite a bit harder to love.

THE fear of the number 13 is known as triskaidekaphobia. The fact that the Last Supper was attended by 13 people appears to have retrofitted to explain the fear. Ditto for the arrest of the Knights Templar on Friday, October 13, 1307. But it bears mention that it is *Tuesday* the 13th that is considered unlucky in Greece, Spain, and a few other cultures.

FRANKLIN Delano Roosevelt was perhaps history's most prominent triskaidekaphobe. It is said that when luncheon or dinner parties numbered 13, he would ask his secretary to join the guests to make an even 14. Not that he invented that particular practice. In France there's even a word for it: A *quatorzieme* is a professional fourteenth guest. And legend has it that Mark Twain found out he was to be the thirteenth guest at a dinner party, at which point a friend told him not to go because it was bad luck. "It was bad luck," Twain later told the friend. "They only had food for 12." But FDR's fear of Friday the 13th—as opposed to, say, fear itself—knew no bounds. In April 1945, as everyone else in America geared up for the Friday the 13th of that month, Roosevelt managed to avoid the day altogether . . . by dying on Thursday the 12th.

NOT every culture hates 13. The Egyptians regarded 13 as a sacred number, and despite FDR's phobia the number has a unique place in American history. Everyone knows about the original 13 colonies as represented by the 13 stars and stripes of the original flag.

But while the flag has undergone numerous revisions (ultimately leaving the number of stripes intact but not the number of stars), if you look at the back of a modern one-dollar bill, you will still find a bunch of thirteens:

13 stars above the eagle
13 steps on the Pyramid
13 letters in ANNUIT COEPTIS
13 letters in E PLURIBUS UNUM
13 vertical bars on the shield
13 horizontal stripes at the top of the shield
13 leaves on the olive branch
13 fruits
13 arrows

▼

ELEVEN PLUS TWO = 13 = TWELVE PLUS ONE

NOT only is the above arithmetic accurate—the three-word expressions to either side of 13 are anagrams of one another.

▼

THE numbers 12 and 13 are also linked by the equations below. The reversal of their squares is also the square of their reversals. (Eleven has this same property rather more trivially, but 13 is the best of all because no digits are duplicated in either the number or its square.)

$12^2 = 144$	$13^2 = 169$
$21^2 = 441$	$31^2 = 961$

▼

THE soccer ball we encountered in **12** is one of 13 Archimedean solids: three-dimensional convex shapes built up of two or more types of regular polygons. Perhaps the most spectacular is the great rhombicosidodecahedron, which is built from 30 squares, 20 hexagons, and 12 decagons.

▼

THE movie *Apollo 13* chronicled the ill-fated 1970 flight of the spacecraft bearing that same name. Like many if not most feature films, *Apollo 13* made a handful of technical errors, including using a logo that didn't appear until 1976, misplacing the Sea of Tranquility, and apparently forgetting that propulsion jets do not make any noise in space. But the most amusing blunder for numbers people came when an engineer at mission control whipped out a slide rule to check the arithmetic of one of the astronauts. The audience laughed at the pure primitivity of it all, drowning out the fact that it was an *addition* problem. The whole point of a slide rule is that numbers are marked in proportion to their logarithms, facilitating multiplication and exponentiation, but not addition.

One postscript: When *Apollo 13* came out on DVD, its release had an unintended effect on the entire movie rental business. A young man named Reed Hastings rented the film and incurred some late fees upon returning it. Upset by what he felt was an unwarranted charge, Hastings founded Netflix, the first DVD-by-mail service.

14 $[\,2 \times 7\,]$

007 × 2

IAN Fleming wrote a total of 14 James Bond novels, one per year from 1953 (the year of his debut with *Casino Royale*) through 1966 (the year of Fleming's death). Two of the 14 books (*Thunderball* and *Octopussy*) were actually collections of short stories.

WE'VE seen that it is impossible to tile the plane with *regular* pentagons, but there are 14 known types of irregular convex pentagons that will tile the plane just fine. Two types are shown below: The first is just a hexagonal tiling with three lines drawn inside each hexagon, while the second, discovered in 1985 by German graduate student Rolf Stein, relies on specific relationships between the sides and angles of the convex pentagon. Stein's creation was the fourteenth pentagonal tiling. The discovery of a fifteenth type would rock the tiling world, but it's a possibility that has yet to be ruled out.

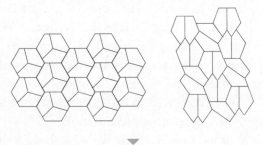

BACH is said to have been especially fond of the number 14, possibly because in the standard alphanumeric code, the letters in BACH add up to 2 + 1 + 3 + 8 = 14.

IN the old British system of weights and measures, a stone equaled 14 pounds avoirdupois.

A *fortnight* is a strange term for a two-week period, until you realize that the word is a contraction of "fourteen nights."

SPEAKING of days, nights, or whatever, there are 14 possible calendars, as January 1 can fall on seven different days, and leap year creates two different calendars for each of those seven choices.

▼

CARBON-14, the basis of the carbon dating technique, is a radioactive isotope whose nucleus consists of 6 protons and 8 neutrons. It occurs naturally and has a half-life of 5,730 years.

▼

IN 1949, the comic duo of Ma and Pa Kettle (Marjorie Main and Percy Kilbride) challenged a salesman's assertion that one-fifth of 25% was 5%, feeling that it was 14% instead. They started by dividing 5 into 25. Well, five won't go into 2, so you have to go to the 5. Five into 5 is 1, so Pa Kettle placed a 1 to the left. What's left when you subtract the 5 from 25 is 20, and 5 goes into 20 four times. Writing a 4 next to the one gave him 14, as he had predicted. And so on, and so on. (Alas, much better on their little chalkboard and washcloth than on the printed page, but their sheer butchery of the number 14 was too thorough not to mention.)

▼

A sonnet is a poem with 14 lines, written in iambic pentameter and separated into an octet and a sestet.

▼

AND let's not forget the Palimpsest. A strange name for a puzzle, but the other names it has picked up are no less strange: the Loculum of Archimedes, or simply the Stomachion. By any name, this is perhaps the world's oldest puzzle. It consists of 14 triangles and quadrilaterals that can be rearranged to form a variety of shapes. One of those shapes is a square, pretty much by definition, because the 14 pieces of the Stomachion (my preferred name, as it essentially means that the solver

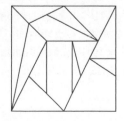

is going crazy) were originally made by cutting along line segments ending on the lattice points of a 12 × 12 square. Since the time of Archimedes (circa 200 BC), the biggest challenge surrounding the puzzle is to compute the total number of ways in which the 14 pieces can be configured to form a square. The answer—536—was not computed once and for all until 2003. And we thought Fermat's Last Theorem took a long time to solve.

15 $[\,3 \times 5\,]$

IN a game of backgammon, each player starts off with 15 pieces, arranged in columns of 2, 3, 5, and 5, as at left.

15 is the fifth triangular number, easily depicted by the pre-break arrangement of 15 billiard balls—in this case, held within the triangular wooden rack used in American eight ball.

GIVING a 15% tip is a common guideline when dining out. And it's a simple matter to calculate 15% tips in your head. Just take 10% of the price of the meal and add half again as much. A $70 check produces a tip of $7.00 + $3.50 = $10.50.

THE 15-letter word *uncopyrightable* is the longest word in the English language that does not repeat any letters.

▼

THE Fifteen Puzzle, supposedly of puzzlemaster Sam Loyd, was a puzzle craze of 1880, almost exactly 100 years before the introduction of Rubik's Cube. The puzzle features square blocks numbered 1 to 15, with the 14 and 15 reversed. The idea is to use the puzzle's one empty square to slide the blocks so as to place the 14 and 15 into their proper positions.

One of the pieces of lore surrounding this puzzle is that a Massachusetts dentist named Charles Pevey offered a reward of a $25 set of teeth (and, soon afterward, $100 in cash) for anyone who could solve the puzzle. Presumably Pevey knew that his money was safe, as no solution is possible. The impossibility of a solution involves what is called a parity argument: Curiously, no matter how the puzzle pieces are arranged, the sum of the number of pieces that are in reverse order plus the row number of the empty square does not change. Because a 14–15 switch would reduce this number by 1, it is unattainable.

Although Sam Loyd took credit for the invention of this puzzle, the actual inventor was apparently a New York postmaster named Noyes Chapman.

▼

8	1	6
3	5	7
4	9	2

THE above figure is said to be a magic square because every row, column, and diagonal sums to the same number. Fifteen is the magic constant for any 3 × 3 square, because the sum of 1 through 9 equals 45 and $\frac{45}{3} = 15$.

▼

IN the discussion in **4**, we saw that any positive integer can be written as the sum of four perfect squares. In other words, any positive integer can be written as $w^2 + x^2 + y^2 + z^2$ for some w, x, y, and z, not necessarily distinct. The number 15 played a role in that discussion because it is the smallest number that requires the full four squares: $9 + 4 + 1 + 1$. But mathematicians can't let a theorem like Lagrange's Four-Square Theorem go without trying to generalize it. And what does such a generalization look like? Well, is it true, for example, that any positive integer can be expressed in the form $w^2 + 2x^2 + 3y^2 + 4z^2$? Let's try it out:

$1 = 1^2 + 2 \times 0^2 + 3 \times 0^2 + 4 \times 0^2$

$2 = 0^2 + 2 \times 1^2 + 3 \times 0^2 + 4 \times 0^2$

$3 = 1^2 + 2 \times 1^2 + 3 \times 0^2 + 4 \times 0^2$

$4 = 2^2 + 2 \times 0^2 + 3 \times 0^2 + 4 \times 0^2$

$5 = 1^2 + 2 \times 0^2 + 3 \times 0^2 + 4 \times 1^2$

$6 = 2^2 + 2 \times 1^2 + 3 \times 0^2 + 4 \times 0^2$

$7 = 2^2 + 2 \times 0^2 + 3 \times 1^2 + 4 \times 0^2$

$8 = 0^2 + 2 \times 2^2 + 3 \times 0^2 + 4 \times 0^2$

$9 = 3^2 + 2 \times 0^2 + 3 \times 0^2 + 4 \times 0^2$

$10 = 1^2 + 2 \times 1^2 + 3 \times 1^2 + 4 \times 1^2$

$11 = 0^2 + 2 \times 1^2 + 3 \times 1^2 + 4 \times 1^2$

$12 = 0^2 + 2 \times 0^2 + 3 \times 2^2 + 4 \times 0^2$

$13 = 1^2 + 2 \times 0^2 + 3 \times 2^2 + 4 \times 0^2$

$14 = 0^2 + 2 \times 1^2 + 3 \times 2^2 + 4 \times 0^2$

$15 = 1^2 + 2 \times 1^2 + 3 \times 2^2 + 4 \times 0^2$

I know this is getting awfully tedious, but, believe it or not, we don't have to go any further. According to an extraordinary 1993 theorem by John Conway and William Schneeberger, any positive-definite quadratic form that represents the first 15 integers will represent any integer whatsoever!

As it turns out, there are 54 expressions of the form $Aw^2 + Bx^2 + Cy^2 + Dz^2$ that represent all integers, ranging from $w^2 + x^2 + y^2 + z^2$ to $w^2 + 2x^2 + 5y^2 + 10z^2$. This list was first identified by the great Indian mathematician Ramanujan in the early twentieth century—that is, unassisted by computers. The only flaw in Ramanujan's work was that he included the form $w^2 + 2x^2 + 5y^2 + 5z^2$, which turns out not to be universal. In fact, the number 15 is the first (and only!) number that cannot be represented as $w^2 + 2x^2 + 5y^2 + 5z^2$ with w, x, y, and z integers.

▼

In the future, everyone will be world-famous for 15 minutes.
—Andy Warhol, 1968

16 [2⁴]

BECAUSE 16 is a perfect square, it is possible to arrange 16 circles in a square, as follows:

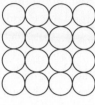

IT is also possible to arrange these same 16 circles in a different sort of square formation:

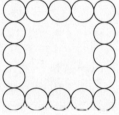

The above square works out because 16, in addition to being a square, is a difference of squares that are two apart, namely $5^2 - 3^2$ or $25 - 9$. (The second figure is just a 5×5 square with its 3×3 interior taken out.) No other number of circles (or whatever) can be configured as two squares in this fashion.

THE equation $16 = 2^4 = 4^2$ is unique. No other number can be represented in the form a^b and b^a with $a \neq b$.

THE fact that there are 16 ounces in a pound is related to 16 being a power of two, sort of. The word for ounce comes from the Latin *uncia*, or twelfth part, making it a cognate of *inch*. The Roman pound was indeed 12 ounces, a standard that remains today in the form of the Troy pound used in the measurement of gold. In between, it seems that seventeenth-century Scottish goldsmiths used a standard of 16 ounces per pound, and, more generally, merchants in the Middle Ages saw the advantage of a unit that could be halved repeatedly, and it was apparently on that basis that today's 16-ounce avoirdupois pound emerged once and for all.

▼

THE hexadecimal base used in computing and other applications is nothing more than base 16, with digits 0, 1, 2, 3, 4, 5, 6, 7, 8, 9, A, B, C, D, E, and F. Because $16 = 2^4$, a hexadecimal bit essentially substitutes for four ordinary bits, making computer codes that much less cumbersome. In Boolean logic, there are 16 possible Boolean operations that can be performed on two variables P and Q: ZERO, P, NOT P, NOT Q, P AND Q, P AND (NOT Q), Q AND (NOT P), and so on.

▼

THERE are 16 basic personality types in the Myers-Briggs classification system, devised by Katharine Cook Briggs and her daughter Isabel Briggs Myers in accordance with ideas published by Carl Jung in 1921. People are either introverted or extraverted (I or E), sensing or intuitive (S or N), thinking or feeling (T or F), and judging or perceiving (J or P). With four basic markers, each having two possibilities, the number of possible combinations is $2^4 = 16$. They are commonly referred to by a set of four letters, as in the following:

ISTJ	ISFJ	INFJ	INTJ
ISTP	ISFP	INFP	INTP
ESTP	ESFP	ENFP	ENTP
ESTJ	ESFJ	ENFJ	ENTJ

The most common personality type is ISFJ—introverted, sensing, feeling, and judging—which accounts for an estimated 13.8% of the population. The least common type, at an estimated 1.8% of the population, is ENTJ—extraverted, intuitive, thinking, and judging. Note that the most common and the rarest types are opposite in three, but not four, of the basic categories.

▼

MOVING to a different type of identification process, for much of the twentieth century, fingerprint identification in the United Kingdom was based on 16 points of similarity. This standard was eventually abandoned because of technological improvements, but the delicious part of the story is that the paper on which the 16-point standard was originally based turned out to have been a forgery.

▼

THIS arrangement of sixteen 0's and 1's is known as a de Bruijn cycle. If you start at the top center of the diagram, the four characters going clockwise form the set {0,0,0,0}. Starting at the next character to the right and going clockwise four spaces yields the set {0,0,0,1}, and so on. By the time you go around the whole circle you will have created each of the 16 possible arrangements of four binary (i.e., either 0 or 1) characters. The existence of de Bruijn cycles for any desired alphabet and cycle size can be proved using graph theory, and these cycles are related to a problem known as universal coloring. This book doesn't happen to use four-color printing, but the related challenge to the above diagram, which you are free to ponder, would be to color each of the 16 positions with one of four colors in such a way that as you went around the circle, you'd come across each of the 16 possible ordered pairs of those four colors—that is, (red, blue), (blue, yellow), and so on.

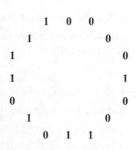

▼

SPEAKING of books, it isn't widely known that the traditional book assembly process uses "signatures" of length 16, which is why the total page count of so many books is something like 272, 288, or some other multiple of 16.

▼

DEFINE two sets of numbers A and B as follows:

$$A = \{1, 4, 6, 7, 10, 11, 13, 16\}$$
$$B = \{2, 3, 5, 8, 9, 12, 14, 15\}$$

It is obvious at a glance that the sets A and B are (1) disjoint and (2) together account for every positive integer from 1 to 16. A second glance reveals that each of the eight pairs {1, 2} through {15, 16} has precisely one element in A and one in B, with four even numbers and four odd numbers in each set, so that the sum of the members of A equals the sum of the members of B. But what is considerably less obvious is that the sum of the *squares* of the elements of A equals the sum of the squares of the elements of B, and similarly for *cubes*. This construction is remarkable but turns out to be possible for any power of 2: a 32-number construction uses fourth powers, a 64-number construction incorporates fifth powers, and so on.

THE 16-square-unit diagram below shows how to create the geometrical shapes known as tangrams. The 7 tangrams—5 triangles, 1 square, and 1 parallelogram—can be combined to create a variety of shapes, including the 4 × 4 square from which they originated.

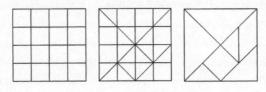

AS an exercise, try dividing the 16-square-unit figure below into the 7 tangram shapes:

OUR final 16-square-unit diagram is the Bachet square, in which each row, column, and diagonal has one ace, one king, one queen, and one jack, with each of the four suits also represented once and only once. These squares were discovered by Claude Gaspar Bachet de Meziriac in the 1600s, and in 1624 he posed the question of how many different ones there were. The answer, 1152, is better recognized as $2 \times 576 = 2 \times (4!)^2$, the idea being that once you fix the top row, there are only two Bachet squares associated with that row, and the rank and suit requirements each end up contributing 4! possible arrangements.

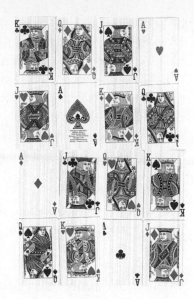

17 [prime]

THE Pythagoreans supposedly detested the number 17, feeling that it was no match for the symmetric beauty of 16 to the left and 18 to the right. They're entitled to their opinion, of course, but history has shown them to be rather foolish in a variety of areas and this one is no exception. Of all the numbers we will encounter in this book, the number 17 is perhaps the biggest surprise in terms of just how much it has going for it.

▼

GEORGE Balanchine had no difficulty finding symmetry in the number 17. That's the number of ballerinas who showed up for one particular class, and Balanchine responded by arranging them in the double-diamond formation, which became the opening for his signature ballet, *Serenade*.

▼

IN Italy, the number 17 takes on the role of bad luck and superstition that 13 occupies in other cultures. Alitalia has no seventeenth row, many Italian buildings do not have a seventeenth floor, and when the Renault R17 went to Italy, its name was changed to R117. This cultural aversion to 17 has long roots, apparently tracing to the anagrammatical relationship between the characters for XVII, the Roman numeral designation for 17, and the Latin word VIXI, whose translation "I lived" somehow morphed into "I am dead." (Note that VI + XI = 6 + 11 = 17.)

▼

THE number 17 is a threshold of sorts for consecutive number sequences. But we'll start at the beginning:

▼

ANY two consecutive integers, pretty much by definition, contain no common factor. They are, in math-speak, relatively prime. That doesn't have to be true of three consecutive numbers, as two of them may be even and thus both divisible by two. But the middle one must be relatively prime to the other two. And among four consecutive integers, one of the middle two must be odd, and is relatively prime to the other three.

How far can we go? You guessed it, any sequence of *fewer than* 17 consecutive integers must contain at least one number that is relatively prime to all the others. But if you look at the 17 consecutive numbers 2184, 2185, 2186 . . . 2199, 2200, you will see that each shares a factor in common with at least one other member of the sequence. In fact, if you choose any number $n \geq 17$, it is always possible to locate a sequence of n consecutive integers such that each shares a factor with at least one of the others. The sequence beginning with 2184 just happens to be the smallest sequence of its kind.

▼

IF you visit Braunschweig, Germany, the birthplace of Carl Friedrich Gauss, you just might come across a statue of Gauss atop a circular pedestal. But give that pedestal a second look. If you do, you'll see that it is not circular but instead forms a figure with 17 equal sides. The statue recognizes one of

Gauss's great early achievements—the creation of a regular 17-gon with only a ruler and a compass (the grade-school compass with a point on one end and a pencil on the other, not a compass that will tell you where the North Pole is).

Let's take a moment to ponder what such a construction is all about. If you were asked to construct an equilateral triangle using only a ruler and compass, you'd proceed as follows: Pick two points A and B. Digging your compass into point A, draw an arc that passes through point B. Now dig the compass into point B and draw an arc that passes through point A. If the arcs you drew are big enough, they intersect at a third point, which we'll cleverly label point C. Now you can finally use your ruler, drawing segments that connect A to B, B to C, and C to A. The result is an equilateral triangle.

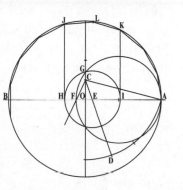

While the 17-gon construction is necessarily complex, the diagram to the right gives a hint of its underlying methods.

GAUSS didn't stop there—as well he shouldn't have, because he constructed the regular 17-gon when he was only 18 years old. He eventually gave an explicit categorization of those regular polygons that lent themselves to a ruler-and-compass construction, along the way proving that it was impossible to create a regular nonagon (9 sides) with such limited means.

THE Cincinnati Zoo has a display that looks much like a 17-gon, except that it displays the various broods of the 17-year cicada, an insect whose extraordinarily annoying chirping noises are made at least tolerable by the fact that they only appear in a given locale once every 17 years. The diagram begins with the appearance of one particular brood in 1987, and ends just before the reappearance of that same brood 17 years later, in 2004.

RECALL from **6** that if you join six dots with lines colored either red or blue, you will automatically create a monochromatic triangle—an all-red or all-blue triangle whose vertices are three of the six dots. If you up the ante to three colors, it turns out that 17 is the magic number.

WE'VE all seen wallpaper patterns and taken note that they repeat in a symmetric fashion. While a wallpaper store or catalog will offer thousands of choices, it might surprise you to learn that any symmetric pattern is one of 17 basic types, appropriately called wallpaper groups. The designs below embody the various symmetry techniques, from translations to reflections to rotations, then on to rotations of different angles and combinations of the above.

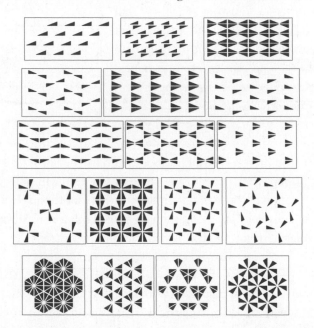

The proof that the number of planar symmetries is limited to 17 is somewhat outside the scope of this book, but there's a great proof in the book *The Symmetries of Things*, by John Conway, Heidi Burgiel, and Chaim Goodman-Strauss.

Scholars appear divided as to whether all 17 planar symmetries are present in the tilings of the Alhambra, the Moorish castle in Granada, Spain (a country that is itself divided into 17 so-called autonomous regions). A brief conversation with David Kelly of Hampshire College, the world's leading authority on the number 17 (no, I'm not kidding) suggests that the answer is yes, though there are some written accounts that are less sanguine. Whatever the final verdict on Alhambra tilings, their reputation inspired an obscure Dutch graphic artist named M. C. Escher to pay visits in both 1922 and 1936. These visits were the cornerstone of Escher's now-legendary pursuit of magical planar symmetries.

▼

FINALLY, consider the Sudoku puzzle below. You may have noticed that there are fewer starting numbers than your typical Sudoku. But if you're thinking that you've seen a published puzzle with fewer givens, there are a lot of mathematicians who'd like very much to talk to you. You see, this puzzle has precisely 17 givens, the minimum number ever achieved at this writing. In other words, no puzzle with 16 or fewer givens has ever been shown to produce a unique solution, and circumstantial evidence is building in favor of 17 remaining minimal, although there is currently no theorem to that effect. Surprisingly, this particular puzzle isn't all that hard. (See Answers.)

	1	4				8		
				2	7			
7	6						2	
		4						
2								
3	7			8				
			5			4		1
						5		

▼

I think I'll stop here as far as 17 is concerned. I'm not sure I've written anything that David Kelly didn't already know, but I trust I've shown enough to demonstrate that the Pythagoreans' dismissal of this wonderful number was woefully off the mark.

18 $\left[2 \times 3^2 \right]$

A rectangle with a perimeter of $3 + 6 + 3 + 6 = 18$ has an area of $3 \times 6 = 18$. Other than a 4×4 square, this is the only rectangle whose area is numerically equal to its perimeter.

▼

STARTING with a single dot placed somewhere on a line segment, it is a simple matter to place another dot so that each dot belongs to its own half of the segment, as below.

Now we place a third dot so that each of the three dots is in its own third. Oops, wait a minute. These two are already in the middle third of the segment. Let's start over.

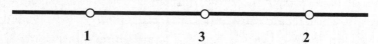

There, that was easy. The first two dots are in different halves, the three dots in different thirds. Can we keep going? Maybe, but only so far. In 1970, Elwyn Berlekamp and Ronald Graham proved the remarkable result that no matter how many times you start over, it is impossible to place 18 points in this manner.

THE 18-Point Theorem, as it is called, has an interesting application in the area of political representation, first written up in 1984 by Virginia Tech economics professor Amoz Kats. Professor Kats's thesis has particular resonance in such places as Israel and the Scandinavian countries, where a legislative body is elected using proportional representation: If a party receives $x\%$ of the votes, it gets $x\%$ of the legislature, with the seats filled in order from the party's list. The question is, can the various constituencies of a party always be represented? In Kats's construct, "A list truly represents the constituency of the party if whenever the first k members of it are elected, each of the k equally spaced sections of the party is represented in the elected body." There's no problem for small legislatures, but the 18-Point Theorem rears its head with larger ones: Kats's specific conclusion was that "a party can construct an *ordered* representative list . . . if and only if it contains not more than seventeen names." Fortunately, this condition is satisfied most of the time, whether or not the party pays homage to the 18-Point Theorem.

THE modern golf course has 18 holes. Volleyball is played on a court that is 18 meters long, with a ball that is divided into 18 sections.

THE number 18 is a standard of sorts for a completely different sport: barrel jumping. As in how many barrels can you jump over with a running start . . . on skates? In the 1960s, the ABC television program *Wide World of Sports* often featured barrel jumping competitions from the Grossinger's resort in the Catskills. The world record of 18 barrels (translating into 29 feet, 5 inches) was set by Yvan Jolin of Canada in 1981, and it's a record that could be around for a long, long time.

AND 18 forms a different barrier still in horse racing. At this writing, the Kentucky Derby winner with the longest name is the 16-letter champion of 2000,

Fusaichi Pegasus (in horse racing, spaces count). The Steven Spielberg co-owned Atswhatimtalknbout raced in 2003, finishing fourth but tying a record that will never be broken. There are 18 letters in Atswhatimtalknbout, and the official rules for the Kentucky Derby (and in many horse racing venues throughout the world) prohibit horse names from exceeding 18 letters.

19 [prime]

THE game Go features a 19 × 19 grid onto which players place white and black pieces in hopes of surrounding their opponent's pieces and thereby removing them from the board. Legends have it that the game dates back to 2000 BC or so.

COMPARED to Go, cribbage is a newcomer on the gaming scene. The invention of cribbage has been attributed to the English poet Sir John Suckling (1609–1642). The game consists of dealing and valuing hands of cards and then moving pegs around a board in accordance with the hand values. It turns out that 19 is an impossible score in the game of cribbage, and is the smallest number with that property. A 19 has come to mean a useless hand.

19 is a so-called centered hexagonal number, meaning that it is possible to place 19 dots in such a way as to form concentric hexagons, with one of the dots forming the center. In the figure below, the numbers 1 through 19 have been placed in the dots so that the sum of the three numbers in any leg of the six triangles is the same: 22. As it happens, there are many such solutions, with a variety of possible sums. The biggest possible sum is 31. Can you find a solution that yields it? (See Answers.)

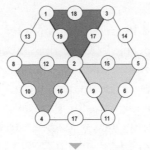

TAKE away the numbers from the above diagram and you're left with the 19-hole hexagonal pattern below, most commonly seen in sink or lavatory drains.

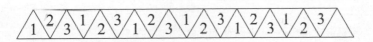

A different sort of association between the number 19 and hexagons starts with the figure above—19 equilateral triangles that can be colored and folded into a hexahexaflexagon. If each of the numbers 1, 2, and 3 is given a different color (and likewise for the back), it is possible by appropriate folding and manipulation to make hexagons of each of the six colors emerge, with none of the other colors visible.

KELLOGG'S Product 19 was introduced in 1967 as a competitive response to Total from General Mills. As the story goes, Kellogg's had trouble coming up with a name for the new product and eventually settled on Product 19, as it was the nineteenth product in company history, Corn Flakes being number 1.

▼

THE philosopher Bertrand Russell once pondered the futility of identifying "the least integer not nameable in fewer than nineteen syllables." By way of comparison, 11 requires three syllables to name, either by saying "e-le-ven" or "eight plus three," and it is the smallest integer with that property. However, Russell was on a completely different wavelength. His point was that the description inside the quotation marks consists of 18 syllables, meaning that the least integer not nameable in fewer than 19 syllables can in fact be named in 18 syllables, a contradiction. Russell attributed this paradox to Oxford University librarian G. Berry. It survives in many different forms, but the basic idea is called the Berry paradox.

▼

QUICK: What does the equation below have to do with the number 19?

$$559 = 256 + 256 + 16 + 16 + 1 + 1 + 1 + 1 + 1 + 1 + 1 + 1 + 1 + 1 + 1 + 1 + 1 + 1 + 1$$

The answer is that there are 19 numbers (summands) on the right side. But there's more. We can rewrite the equation as:

$$559 = 4^4 + 4^4 + 2^4 + 2^4 + 1^4 + 1^4 + 1^4 + 1^4 + 1^4 + 1^4 + 1^4 + 1^4 + 1^4 + 1^4 + 1^4 + 1^4 + 1^4 + 1^4 + 1^4$$

In other words, the number 559 can be written as the sum of 19 fourth powers. But guess what? *Every single positive integer* can be written as the sum of at most 19 powers. Perhaps you remember our discussion in **4** concerning Lagrange's Theorem, which states that every positive integer can be written as the sum of four squares. Well, mathematicians didn't leave that one alone. In 1770, Oxford mathematician Edward Waring conjectured among other things that 19 fourth powers would suffice to represent any positive integer. The result was finally proved in 1986, by Balasubraman-ian, Deshouillers, and Dress. It turns out that only seven numbers (559 being the largest of the seven) require the full 19.

▼

1	1	2	3	5	8	13	21	34	55	89	144	233	377	610	987	1597	2584	4181
F_1	F_2	F_3	F_4	F_5	F_6	F_7	F_8	F_9	F_{10}	F_{11}	F_{12}	F_{13}	F_{14}	F_{15}	F_{16}	F_{17}	F_{18}	F_{19}

ABOVE are the first 19 Fibonacci numbers, where $F_1 = F_2 = 1$ and each suc-cessive member of the sequence is obtained by summing the previous two. (See **5**, **8**, etc.) The entries that are highlighted in gray are those Fibonacci numbers with *prime* subscripts (not including F_2). The first six prime sub-scripts—3, 5, 7, 11, 13, and 17—each produce a Fibonacci number that is prime. There is no good reason why this pattern should continue, and in fact F_{19} is the first of infinitely many exceptions: F_{19} equals 4,181 = 37 × 113.

▼

SPEAKING of divisibility, any number that is divisible by 19 will have the fol-lowing peculiar property: If you multiply the last digit by 2 and add that number to the remaining digits, the resulting number will be divisible by 19. For example, 625632 is divisible by 19 because all of following results are divisible by 19: (62563 + 4 = 62567); (6256 + 14 = 6270); (627 + 0 = 627); (62 + 14 = 76); (7 + 12 = 19); (1 + 18 = 19). As a divisibility test, this procedure is relatively easy to describe, but it usually requires so many it-erations that you may wonder why you didn't just divide by 19 in the first place. But don't take my word for it. If you like, you can try the process out with one of the most famous multiples of 19, the number 19181716151413121110987654321—formed by stringing together the first 19 integers in reverse order.

▼

IN general, 19 divides a positive integer if and only if (abbreviated iff) 19 di-vides the number that results from adding twice the value of the last digit that results from stripping off this last digit. For example, $19|704836$ iff $19|70495$ iff $19|7059$ iff $19|723$ iff $19|78$ iff $19|23$ iff $19|8$. Since this last divisibility $19|8$ is obviously false, 19 does not divide evenly into 704836.

20 $\left[\, 2^2 \times 5 \,\right]$

THERE are 20 possible first moves for either player in a game of chess: Any of the eight pawns can be moved forward either one or two squares, while either of the two knights can be moved two squares up and one square either to the left or to the right. (In standard chess notation, the possible moves for white are a3, a4, b3, b4, c3, c4, d3, d4, e3, e4, f3, f4, g3, g4, h3, h4, Na3, Nc3, Nf3, and Nh3. Four of the 20 possible moves are made by a knight—labeled N because K is already taken for king—while the other 16 moves are made by pawns, so lowly that modern notation doesn't even give them a letter.)

▼

THE number 20 shows up elsewhere in the worlds of games and music, notably in 20 Questions.

▼

WHILE Dungeons & Dragons uses a 20-sided die (an icosahedron) that everyone can see, the Magic 8-Ball, invented by Abe Bookman in 1946, relied on an icosahedron suspended in blue fluid inside the ball. You ask the ball a question, shake it up, and wait for the answer to emerge in the window on the ball. The 20 standard answers are as follows:

Signs point to yes.	Yes.
Reply hazy, try again.	Without a doubt.
My sources say no.	As I see it, yes.
You may rely on it.	Outlook not so good.

Concentrate and ask again.	It is decidedly so.
Better not tell you now.	Very doubtful.
Yes—definitely.	It is certain.
Cannot predict now.	Most likely.
Ask again later.	My reply is no.
Outlook good.	Don't count on it.

AN even sneakier icosahedral structure is at the heart of many viruses, as first conjectured in 1956 by Francis Crick and James Watson of DNA fame.

THE number 20 has always been considered a viable number for a numerical base, because it coincides with the total number of digits in the human body. In the former British currency system, 20 shillings made up a pound.

IN the non-metric world, 20/20 vision is the standard for normal vision, literally meaning that what you see at 20 feet is what you should normally see—as opposed to, say, 20/60 vision, in which you see at 20 feet what a person with normal eyesight sees at 60. In the metric world, this same concept is often described as 6/6 vision, where the 6 stands for meters.

As desirable as 20/20 vision is, someone described as applying 20/20 hindsight is seldom being lauded. Apparently to be a true visionary you have to make decisions before you know how they turn out.

21 [3 × 7]

THE factorization of the number 21 into two prime factors was apparently not lost on Franklin Roosevelt. When Roosevelt took office in 1933, one of the president's regular tasks was to set the price of gold. Sounds weird, doesn't it? It sounds even weirder when you hear that on one occasion Roosevelt proposed raising the gold price by 21 cents . . . on the grounds that 21 was three times seven and therefore a lucky number. (One can presume that Roosevelt never raised the gold price by 13 cents, given his legendary fearfulness of that number. See **13**.) Actually, the number 21 really did turn out to be lucky within the Roosevelt administration, in the sense that the 21st Amendment to the Constitution, ratified in December 1933, repealed Prohibition. Today, of course, it is legal to drink alcohol in any of the 50 states—provided, of course, that you are at least 21.

▼

THE game 21, otherwise known as blackjack, is so called because a gambler seeks to come as close to 21 points (an ace and a face card, for example) without going over. This same idea was used in the infamous 1950s game show *Twenty One*, in which participants would get points for answering questions, with the option of stopping if they felt they were closer to the magical total of 21 points than their opponent was. Some of the participants were prodigiously smart with their answers and uncannily accurate in playing the 21 game, but within a couple of years it was revealed that the entire show was staged, as were many other quiz shows during that era. The 1994 movie *Quiz Show* focused specifically on *Twenty One*. As for the movie *21*, well, that focused on blackjack and was based only in loose fashion on Ben Mezrich's book *Bringing Down the House*, the tale of some MIT undergraduates who took Las Vegas for a ride by using a card-counting system.

Quiz Show and *21* offer a pair of symmetric flubs to go with their symmetric background. In the former, the film shows Jack Barry as the emcee of *Twenty One* when the quiz-show scandals broke in the summer of 1958, but in fact the emcee that summer was none other than Monty Hall. Yes, that's the same Monty Hall of *Let's Make a Deal* fame, now immortalized mathematically by the Monty Hall paradox, which we introduced in **3**, and *21* showcased in its early stages, when the professor played by Kevin Spacey gave the problem to Ben Campbell, star student and future star gambler. While Ben got the paradox right, noting that the chance of winning had gone from 33.3% to 66.7% (see **3**), the awkward use of decimals and percentages felt like a director's decision to spare his audience the ickiness of fractions. In the words of David Boyum, coauthor with yours truly of the quantitative reasoning guide *What the Numbers Say*, "There is a 0.0% chance that an MIT math whiz would say 33.3% and 66.7% instead of the correct $\frac{1}{3}$ and $\frac{2}{3}$." At least *21* tried to redeem itself by putting the sequence 1, 1, 2, 3, 5, 8, 13, . . . on Ben's twenty-first birthday cake, well aware that 21 was the next number in the Fibonacci sequence.

THE number 21 shows up elsewhere in the world of games. A game of table tennis is won by the first player to 21 points. The same was true in horseshoes until 1982, when the 21-point game was officially shelved in favor of a 40-point game. Twenty-one is also the total number of dots on a standard die, being the sum of the numbers 1 through 6. In mathematical terms, 21 = T_6, the sixth triangular number.

IN any given year, the National Hurricane Center generates 21 official hurricane names for the storm season.

In general, people named Quentin, Upton, Xavier, Yvonne, and Zelda never have to worry about having a hurricane named for them, because they represent the five letters that the National Hurricane Center omits on the grounds that they don't generate enough names.

IT'S easy to subdivide a square into four smaller squares, but if you stipulate that no two smaller squares can be the same size, you need a minimum of 21 squares to accomplish the dissection. It turns out that there is only one such minimal dissection (subject to rotations and reflections), and that is the one given here. It was discovered by A. J. W. Duijvestijn in 1978.

22 [2 × 11]

A representation of a positive integer as the sum of positive integers is known as a partition. The number 22 shows up in several places in partition theory. For example, there are exactly 22 partitions of the number 8, starting with $1 + 1 + 1 + 1 + 1 + 1 + 1 + 1$ and ending with 8 itself. Perhaps more interesting is the following pair of partitions of 22:

$$22 = 4 + 5 + 6 + 7$$
$$22 = 1 + 4 + 7 + 10$$

In the top line, the four numbers being added are one apart. In the lower addition, they are three apart. The number 22 is the smallest number that can be written as the sum of evenly spaced integers in two different ways.

Even more interesting is the following trio of partitions of 22:

$$22 = 3 + 3 + 4 + 12$$
$$22 = 2 + 5 + 5 + 12$$
$$22 = 2 + 4 + 8 + 8$$

In each case, the sum of the reciprocals of the members of the partition equals 1: $\frac{1}{3} + \frac{1}{3} + \frac{1}{4} + \frac{1}{12} = 1$, and similarly for the others. Partitions of this sort are sometimes called exact partitions, and 22 is the smallest number with more than one exact partition.

▼

ASIDE from number theory, partitions also play a role in the analysis of systems of particles. Energy levels for individual particles appear as exponents, which are then added when calculating the energy of a thermodynamically closed system. A "partition function" means something different in this context, but the underlying question can still be the representation of positive integers by a bunch of smaller ones.

▼

TURNING to different kinds of partitions, whereas Gaul was renowned for having precisely three parts, modern-day France is partitioned into 22 regions, from Alsace to Upper Normandy.

AS it happens, 22 is also the maximum number of pieces that can be created out of six intersecting lines or five intersecting circles, as shown in the diagrams below. Somehow neither diagram looks much like France. It is inevitable in such constructions (at least, the ones I draw) that certain regions end up much bigger than others. Note that the twenty-second region of the right-hand drawing is the area *outside* all of the five circles. In general, n lines can divide the plane into as many as $\frac{1}{2}(n^2 + n + 2)$ parts, while n circles can divide the plane into as many as $n^2 - n + 2$ parts.

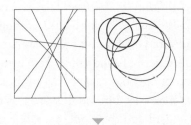

IN elementary mathematics, the best-known appearance of the number 22 is as the numerator of $\frac{22}{7}$, the most commonly used approximation for π. But get a load of this: In the United States, a man's hat size of 7 actually corresponds to a woman's size of 22, because the size of a woman's hat is generally given as the circumference of the inner sweat band, while the size of a man's hat is the diameter of that band, if reconfigured to form a perfect circle. The correspondence of 22 and 7 is then to be expected, as π is by definition equal to the circumference of a circle divided by its diameter.

THE diagram on the next page uses only the numbers 1 through 22 and has the remarkable property that the sum of any two numbers joined by a line segment equals a prime number.

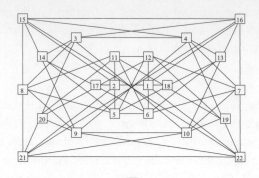

SINCE Joseph Heller's breakthrough novel of 1961, *Catch-22* may be the most recognizable use of the number 22 in modern culture. The expression *catch-22* is nonetheless disappointing for the utter lack of 22-ness to it. Apparently Heller's book was originally called *Catch-18* before being re-titled in deference to Leon Uris's *Mila 18*, a novel about the Warsaw Uprising that also came out in 1961. Numbers such as 11, and 14, and 17 were eventually proposed and rejected. No matter. In the book *Catch-22*, only crazy pilots had a chance to avoid certain combat missions, but mere awareness of the absurd dangers of those missions was considered the work of a rational mind, not craziness. Today we can use *catch-22* to describe a wide variety of no-win situations.

A cricket pitch measures 22 yards, the same as the length of a Gunter's chain (see **66**) and also one-tenth of a furlong. *Furlong* is a contraction of "furrow long," as these and other early measurements arose in the context of plowing through farmland. A strip of land measuring one furlong by one chain was known in olden days as a Saxon strip-acre, having the same square footage as a modern acre (43,560 square feet) but restricted to those specific rectangular dimensions.

A solid cube Greek cross is formed by putting together five cubes, or alternatively 22 squares.

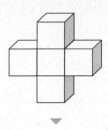

FINALLY, the product of the first 22 positive integers, written 22!, is the 22-digit number 1,124,000,727,777,607,680,000. It turns out that 22, 23, and 24 are the only positive integers *n* (aside from the trivial case *n* = 1) for which *n*! has precisely *n* digits.

23 [prime $\qquad$ $2^3 + 3^2 + 2 \times 3$]

THE number 23 was given an enduring place in the history of mathematics at the dawn of the twentieth century, when German mathematician David Hilbert presented 23 unsolved problems as a challenge to his peers. One measure of the difficulty of these 23 problems is that the easiest of the lot turned out to be this one: Given any two polyhedra of equal volume (such as the cube and tetrahedron below), is it possible to dissect the first into finitely many polyhedral pieces that can then be arranged to form the second?

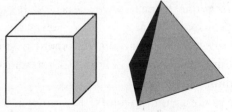

The answer, produced by Hilbert's student Max Dehn within a year, was no. Hilbert suspected this very result, but he was well-aware, as you are now, that such dissections and rearrangements are always possible for polygons—

that is, two-dimensional figures. (The Bolyai-Gerwein Theorem from the early nineteenth century.)

Perhaps the most famous of Hilbert's problems is number 2, which asked if the laws of arithmetic (i.e., Peano's axioms) could be proved to be internally consistent. An answer of sorts came in 1931 in the form of Gödel's Incompleteness Theorem, which shocked the mathematical world by concluding that no system of axioms for arithmetic can be complete. Unlike propositional logic (p implies q and all that), any arithmetical system built on axioms will always produce statements that are true according to those axioms but not provable using those same axioms. So if you want to prove Peano arithmetic consistent, good luck, but you can't do it within Peano arithmetic itself.

The proof of Gödel's Theorem (actually *theorems*, but we'll spare you the distinction among his various efforts) revolved around a coding system for mathematical notation and expressions (called Gödel numbers) and the creation of self-referential statements having paradoxical implications. The Berry Paradox in **19** gave a flavor of how such statements can work. We now consider a conundrum introduced by logician Raymond Smullyan in his Gödel's Incompleteness Theorems of the Oxford Logic Guide series:

Start by imagining an island of knights and knaves, in which knights only make true statements and knaves only make false ones. As Smullyan points out, "No inhabitant can claim that he is not a knight (since a knight would never make such a false claim and a knave would never make such a true claim)." And the plot thickens when a logician—who never believes anything that is false—visits the island and meets a native, who comes out with the surprising declaration, "You will never believe that I'm a knight." I'll let Smullyan tell you what happens next:

"If the native were a knave, then his statement would be false, which would mean that the logician *would* believe that the native is a knight, contrary to the assumption that the logician never believes anything false. Therefore, the native must be a knight. It, then, further follows that the native's statement was true and, hence, the logician can never believe that the native is a knight. Then, since the native really is a knight and the logician believes only true statements, he also will never believe that the native is a knave. And so the logician must remain forever undecided as to whether the native is a knight or a knave."

Gödel's Theorem is just confusing enough to have been misapplied by mathematicians, philosophers, and theologians alike. Swedish logician Torkel Franzen has chronicled and analyzed such nonsense as "Gödel's Theorems show that the Bible is either inconsistent or incomplete"; "By Gödel's Incompleteness Theorem, all information is innately incomplete and self-referential"; or even "By equating existence and consciousness, we can apply Gödel's Incompleteness Theorem to evolution." But Gödel's efforts also led mathematicians to concrete proofs of the undecidability of a variety of propositions in computing, set theory, and even algorithms for the solution of Diophantine equations.

▼

PERHAPS the most difficult of Hilbert's problems was the Riemann Hypothesis. Have you heard of this one? It's gotten some play in the popular press just for the daunting challenge it represents, but technically it's a conjecture about when a function of complex variables called the Riemann zeta function takes on the value zero: The first billion-plus "zeroes" have been shown to lie on a special line in two-dimensional space, but mathematicians won't be happy until a theorem emerges covering *all* zeroes. Although the Riemann Hypothesis is couched in terms of complex numbers (the so-called imaginary numbers of high school, those involving the dreaded $i = \sqrt{-1}$) its primary application is to refine our understanding about the distribution of *prime* numbers. Hilbert seemed to understand the problem's intrinsic thorniness when he uttered the memorable 23 words, "If I were to awaken after having slept for a thousand years, my first question would be: Has the Riemann Hypothesis been proven?"

And Hilbert wasn't alone in his wondering. Those who watched the movie *A Beautiful Mind* may recall a backyard scene in which the protagonist John Nash told a mathematical colleague/visitor that he was working on the Riemann Hypothesis. The camera followed the visitor's eyes to a notebook that contained only the random scribbles of a paranoid schizophrenic, and certainly nothing approaching a solution to one of the holy grails of higher mathematics. The visitor rolled his eyes, and the problem remains unsolved at this writing. Nash, as it happens, had a special fondness for the number 23. One of his most inspired delusions was that *Life* magazine had done a story on him in which he was disguised as Pope John XXIII, in whose pontificate (1958–1963) Nash's schizophrenia first emerged.

▼

HERE'S an odd one: Because we are discussing it here, we know without counting that the following sequence contains 23 letters. But what does it represent? And why are three letters missing? (See Answers.)

O N E T W H R F U I V S X G L Y D A M B Q P C

▼

AND here's an odder one: If you start with a square, you need 23 line segments of equal length in order to make that square rigid—imagine that the square is made of toothpicks, then surrounded by identical toothpicks until its movement is completely restricted. (The symmetry of the drawing is revealed if you rotate it 45 degrees.)

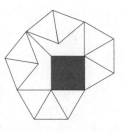

▼

FINALLY, the number 23 plays a decisive role in the so-called birthday paradox. The paradox is the answer to a question: How many people would you have to assemble in order to assure that the probability of some shared birthday among those people is greater than one-half? The answer—just 23 people—seems impossibly small. After all, you'd need 367 people in a room before a shared birthday became 100% certain. (That's an easy application of the pigeonhole principle—see **37.**) The birthday paradox often remains counterintuitive even after an explanation, but I'll give it a try:

The reason that the number is so low is that you're not looking to match the birthdays of any two particular people, nor do you care on what date the shared birthday happens to fall. Any match will do. Imagine placing 23 objects into 365 boxes. There are a whole lot of ways of doing it so that no two objects are placed in the same box (365 × 364 × . . . × 343). Yet there are also a whole lot of ways of placing the objects so that two or more *do* end up in the same box. Start by choosing any two of the objects at random and placing them in box 1, then distribute the remaining 21 objects in the other 363 boxes, and so on, and so on. When all is said and done, 23 objects are enough to create a 50% probability that some box has more than one object in it. The chart below demonstrates that a shared birthday crossed the 50% threshold at 23 and leads to virtual certainty within a group of 60 or more people.

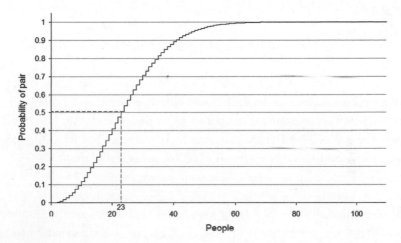

The last word on the birthday paradox belongs to Raymond Smullyan, just introduced as a logician but also skilled in other branches of mathematics. Here's a snippet of a correspondence I received from him while this book was in progress:

I was teaching a course in probability at Princeton, and at one point I told the class that if there are more than 23 people in a room, the chances are more than 50% that at least two of them have the same birthday. I then told them that since there were only 19 students in the room, the chances were extremely small that two of them had the same birthday. One student then said, "I'll bet you a quarter that two of us here have the same birthday!" I thought about this and said, "Oh, of course! You know the birthday of someone here other than your own!" He replied, "I can assure you that I do not know the birthday of anyone here other than my own. Nevertheless I'll bet you that at least two of us here have the same birthday." Well, I thought I would teach him the error of his ways and so I took the bet. I then asked one student after another his birthday, but at one point suddenly realized that two of them were identical twins! Boy, the class had a really good laugh! I then said, "This really shows the futility of pure theory when not backed by empirical observation!"

24 $\left[\ 2^3 \times 3\ \right]$

THE number 24 makes an appearance in some of mankind's oldest games. There are 24 points on a backgammon board, a point being one of those thin triangles that give the game board its distinctive look. And there are 24 dots in the game Nine Men's Morris, which is so old that its fading popularity was noted in Shakespeare's *A Midsummer Night's Dream*.

▼

AND, of course, the number 24 is associated with timekeeping. Everyone knows that there are 24 hours in a day. Not everyone knows that 24 frames per second is a longtime standard in the movie industry.

▼

WHEN you blend sports with timekeeping, you move inexorably to the 24-second clock, a fixture at NBA games since the 1954–55 season. The clock was the brainchild of Danny Biasone, onetime owner of the Syracuse Nationals franchise. The idea was to increase shooting and scoring. Biasone estimated that an average NBA game produced a total of 120 shots, or one shot every 24 seconds for 48 minutes. He was well aware that most possessions would not use up the full 24 seconds, so a 24-second clock would necessarily lead to more shooting and thus more scoring, and that's exactly what happened. The lowest scoring game since the adoption of the shot clock was that very same year—1955—when the Boston Celtics beat the Milwaukee Hawks, 62–57. Despite the emphasis on tight defense in the modern game, the advent of the 3-point shot makes it extremely unlikely that this record will be broken, much less the Pistons–Lakers all-time record of 19–18. (Fort Wayne over Minneapolis, that is, in 1950.)

▼

24 is the product of the first four positive integers, usually written 4! Admittedly, the exclamation point looks silly at the end of the sentence. It's a little bit like writing "Today I watched *Jeopardy!*" If the reader doesn't know that the exclamation point is part of the title, it looks as though you're just really, really excited about having watched a TV show. What I meant was 4!—read as "4 factorial" and meaning the product of the positive integers less than or equal to 4—equals 24.

For any n, n factorial is the product of all positive integers less than or equal to n, and it also happens to equal the number of ways of arranging n objects: n ways to place the first object, $n - 1$ ways to place the second, and so on. In particular, there are 24 ways of arranging 4 objects, illustrated below by using the letters O, P, S, and T. In the layout below, all six entries in the first column are words, and in fact no other choice of four letters will yield more than six words.

OPTS	OPST	OSTP	OSPT
POST	OTSP	OTPS	PSOT
POTS	PSTO	PTOS	PTSO
SPOT	SOPT	SOTP	SPTO
STOP	STPO	TOSP	TPOS
TOPS	TPSO	TSOP	TSPO

▼

THE concept of ordered sets of four letters can be taken a step further, leading to the plausible (but not dictionary-approved) word ANTITRINITARIANIST—someone who opposes the Christian doctrine of the Trinity. To see what makes this "word" special, note that "doctrine" contains the letters

TRIN in that order. "Trinity" starts with TRIN but also contains the sequences RINT and RNIT, though not consecutively. Well, if you look closely enough at ANTITRINITARIANIST, you will find *all* 24 rearrangements of the letters I, N, R, and T imbedded somewhere inside.

The drawing above could never be considered antitrinitarianist. That 24 people are represented is no surprise. But what does the drawing depict? (See Answers.)

▼

24 can be expressed as the product IV × VI, a Roman numeral palindrome.

▼

TO arrive at one of the most remarkable properties of the number 24, start by adding one squared plus two squared. You should get $1 + 4 = 5$. Now add one squared through three squared. You get $1 + 4 + 9 = 14$. Note that neither 5 nor 14 are perfect squares, and there's no reason why they should be. Not until you add the first 24 perfect squares do you get a perfect square: $1^2 + 2^2 + \ldots + 23^2 + 24^2 = 4900 = 70^2$.

The situation above was commonly known in the mathematics literature as the cannonball problem. If you make a square consisting of 24 cannonballs on a side, then place atop those cannonballs an inner square consisting of 23 cannonballs on a side, you can keep going until you make a pyramid, and the total number of cannonballs in the pyramid is a perfect square. The beautiful part is that 24 is not only the first number (other than 1) that makes this construction possible; it is the last.

▼

A cube has 24 rotations: four for each of the six faces. There are also precisely 24 jigsaw-like pieces that can be created by starting with a square and giving each side either a male or female connection or no connection at all.

The bi-sex are seven (one side). I am lucky to find four angles. Of all, these 24 pieces can be interlocked to form a 4 × 6 rectangle.

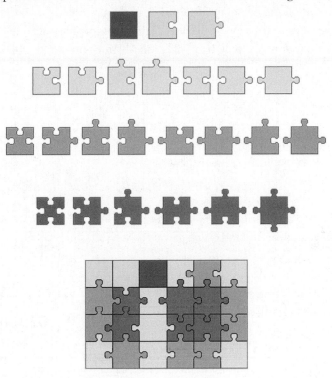

25 [5^2]

BECAUSE the expression 5^2 represents 25 using each of its digits precisely once, 25 is called a Friedman number, in honor of Erich Friedman of Stetson University in DeLand, Florida. All one-digit numbers are trivially Friedman numbers, but 25 is the smallest two-digit Friedman number. It has been conjectured that all powers of 5 are Friedman numbers.

WHEN you take 25 to any power, the resulting number always ends in 25. Such a number is called an automorphic number, and 25 is the smallest two-digit automorphic number. (The one-digit automorphic numbers are 1, 5, and 6.) A number is divisible by 25 if and only if it ends in 25, 50, 75, or 00.

THE number 25 is the smallest square that is the sum of two squares, meaning that it is the square of the hypotenuse in the smallest possible Pythagorean triple: (3, 4, 5). It is also the hypotenuse in another triple (7, 24, 25) featuring consecutive integers.

THE card game 25 is considered the national card game of Ireland. The game Pachisi depicted at right is the national game of India, and *pachisi* is Hindi for "25." Players start and finish in the middle square called the Charkoni. Note that each of the four legs of the game board has 3 × 8 = 24 slots, so you get back to the Charkoni by moving a total of 25 spaces, which happens to be the maximum that can be produced by the rolling of the cowrie shells used in a dice-like fashion (don't ask) to determine movement in the game.

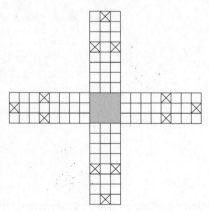

A standard Bingo card has 25 spaces, 24 of which are occupied by numbers.

26 $\left[\, 2 \times 13 \,\right]$

26 is better known for being half of something than something in its own right. Twenty-six weeks make up half a year, and 26 playing cards constitute half a deck, as you can see from the factorization above. Since there are 13 cards in each suit, any two suits together consist of $2 \times 13 = 26$ cards. Alternatively, in a game of bridge each side starts off with a total of 26 cards. It's still half a deck.

▼

THE equation $2 \times 13 = 26$ also shows up in croquet. The standard form of the game is played with two balls, each of which can generate 13 points: 12 hoops and the center peg. The winner of a game of croquet therefore has a total of 26 points.

▼

IF you subtract 1 from 26 you get 25, a perfect square. If you add 1 to 26 you get 27, a perfect cube. No other number is nested between a square and a cube in that fashion.

▼

SPEAKING of proximity, cards, and the number 26, take out a deck of cards. Now think of any two cards in the deck. What do you suppose the probability is that these two cards are next to one another? The question supposes that the cards are thoroughly shuffled, meaning that it's no fair to unwrap a new deck and choose the king and queen of diamonds. Assuming that the cards are in fact randomly distributed, the likelihood that your two cards will be next to one another turns out to be precisely one in 26.

▼

BECAUSE there are 26 letters in the alphabet, the number 26 plays an important role in cryptography. We'll start with an easy encryption to build our confidence. In the sentence below, each letter stands for a different letter of the alphabet:

SGE ZNK LUXIK HK COZN EUA

Unraveling this sort of gibberish becomes easier upon the revelation that the underlying code is an old-fashioned shift cipher, meaning that every letter in the original quotation is shifted ahead some constant number of spaces in the alphabet to produce the coded version. One of the weaknesses in shift ciphers is that the letter appearing most frequently—in this instance, K—is likely to be associated with the most common letter in the English language, namely E. Sure enough, that's the case here. The letter K is six spaces ahead of E in the alphabet, so to break the code you must take every letter back six spaces, obtaining the more familiar:

MAY THE FORCE BE WITH YOU

There's a teensy-weensy wrinkle in setting up the code, and that's where 26 enters the picture. To go from, say, *M* to *S* involves counting six spaces to the right (from position 13 in the alphabet to position 19). That's obvious enough. However, in order to go from *Y* to *E* (the first letter of the last word), you must wrap around from the twenty-fifth letter to the fifth. In mathematical terms, we are using modular arithmetic, a term that somehow complicates a process that is intuitively simple (especially when you're looking at the alphanumeric table below). In modular arithmetic, we say that $25 + 6 = 31$ is congruent to 5 (mod 26), because 31 gives a remainder of 5 upon division by 26. It's the same principle that makes three hours after 11:00 a.m. not 14:00 but 2:00 p.m. The only difference is that clock arithmetic uses a modulus of 12 rather than 26. Again, the use of words like "modulus" makes the process look messier than it really is. The bottom line is that the letter *Y* is encoded by using the letter *E* when using a shift of 6.

A	B	C	D	E	F	G	H	I	J	K	L	M	N	O	P	Q	R	S	T	U	V	W	X	Y	Z
1	2	3	4	5	6	7	8	9	10	11	12	13	14	15	16	17	18	19	20	21	22	23	24	25	26

The whole business of shift ciphers dates back to Julius Caesar, who apparently preferred a left-shift of three spaces—not terribly sophisticated militarily, but remember that the Roman Empire's prior key adversary had ridden elephants through the Alps. A shift cipher of 13 is perhaps the easiest to convey, because you just place the first half of the alphabet on top of the second and switch every vertically aligned pair. Note that under this construction the word VEX doesn't change much, as it is converted into IRE, and vice versa.

There are many different ways of using the above table to form a code. (It's worth pointing out that A is often given the value 0, with Z then equaling 25.) For example, using modular arithmetic, we could *multiply* the values in MAY THE FORCE BE WITH YOU by 3 to get the code MCW HXO RSBIO FO QAHX WSK. Note that M gets mapped into itself by this method, because $3 \times 13 = 39 = 13 \pmod{26}$. But multiplying by 2 wouldn't have worked, for several reasons. Why not? Well, suppose the letter N appears in the code. N has a numeric value of 14, which is 2×7, so it appears that N stands for the seventh letter, G. But why couldn't it stand for T instead? T is the twentieth letter, and multiplying 20 by 2 gives 40, which upon subtracting 26 for the wraparound yields 14, or N.

The general principle working here is that for multiplicative ciphers you can only multiply by numbers that have no common factor with 26, namely the following twelve numbers: 1, 3, 5, 7, 9, 11, 15, 17, 19, 21, 23, and 25. These numbers are said to be relatively prime with 26. (Admittedly, using 1 as a multiplier doesn't generate much of a code.) For a given integer n, the number of integers less than and relatively prime to n is given a name: $\phi(n)$ (read "phi of n"). Also called the totient function, $\phi(n)$ can be calculated by the formula $\phi(n) = (n)\pi (1 - \frac{1}{p})$, indicating a product where p takes on all distinct primes that divide into n. In the case of $n = 26$, 2 and 13 are the only prime factors, so $\phi(26) = 26(1 - \frac{1}{2})(1 - \frac{1}{13}) = 26(\frac{1}{2})(\frac{12}{13}) = 12$.

▼

THERE are a few more English-language tie-ins for the number 26. One is that 26 is the maximum number of words you ever have to create in the game Word Whomp. Word Whomp is a downloadable online game in which players have a limited amount of time (2+ minutes) to create all possible words from a set of six letters, including at least one word using all six of those letters. Sometimes that six-letter word contains very few smaller words, but every word in the makeshift sentence STRONG MAYORS LOOSEN TENDER BOUGHS contains the 26-word maximum. (Perhaps "maximum" belongs in quotes there, as a Scrabble dictionary would surely generate different entries for many words.)

And it's impossible to resist including the sentence "Mr. Jock, TV quiz PhD, bags few lynx." Nonsensical, perhaps, but it's one of the best "sentences" that uses every letter of the alphabet precisely once.

27 [3³]

THE number 27 is not only a perfect cube, but it also bears an unusual relationship with its own cube. Twenty-seven cubed equals 19,683, and if you add the digits in that number you get $1 + 9 + 6 + 8 + 3 = 27$. As it happens, 26 shares this same property, but no larger number does, and in fact the only numbers (other than 1) that do are 8, 17, 18, 26, and 27. How nice that this sequence begins and ends with a perfect cube.

▼

AND that's not the only thing perfect about the number 27. In a perfect game in baseball, a pitcher faces the minimum of 27 batters—3 per inning for 9 innings.

▼

A different sort of appearance of 27 in the world of fun and games is that a three-person game of Rock-Paper-Scissors has a total of 27 possible outcomes. This is of course just another way of saying that 27 equals 3 cubed.

▼

IN New Zealand, Australia, and the United Kingdom, housie cards are 3 × 9 rectangles, with the first column containing numbers between 0 and 9 and so on—the ninth column containing numbers between 80 and 89.

5				49		63	75	80
		28	34		52	66	77	
6	11				59	69		82

▼

IN snooker, the six non-red colored balls have point values 2, 3, 4, 5, 6, and 7 (yellow, green, brown, blue, pink, and black, respectively). Their combined point value is therefore the sum of the numbers 2 through 7: 27. Can you find another two-digit number that equals the sum of the numbers between its first and second digit, inclusive? (See Answers.)

▼

A 27-speed bicycle gets its name because it has three chain-rings and nine cogs, so theoretically you have 3 × 9 combinations. While that's technically true, these chain-ring/cog combinations overlap in terms of distance per crank-arm revolution, the figure that actually determines speed. In practice, a 27-speed bike has something like 15 discernibly different speeds. Credit for this observation goes to David Boyum, coauthor of the quantitative reasoning book *What the Numbers Say*.

▼

27 is the smallest number whose reciprocal has a three-digit repeating decimal: $\frac{1}{27}$ = 0.037037037. . . Curiously, 37 is the next (and only other) number with this property, and $\frac{1}{37}$ = 0.027027027. . . This relationship looks mystical, but all it means is that 27 × 37 = 999. In general, a number has an n-digit repeated decimal expansion if and only if it divides into a string of n 9's and no smaller such string.

28 $\left[2^2 \times 7 \right]$

EVERYONE knows that February is the shortest month, with just 28 days in a non–leap year. One nicety of this length is that 28 days is precisely four weeks, meaning that the day of the week for a date in March will be the same as for the corresponding date in February. Not much of a convenience, but at least it's with us for that one particular month three out of every four years.

HERE'S another nifty trick of the calendar, courtesy of the number 28. As a general rule, any two calendars that differ by 28 years must be identical: With 4 years between leap years and 7 days in a week, the calendar cycle automatically renews itself every $4 \times 7 = 28$ years. Yes, there are exceptions, so please don't write in. For example, T. S. Eliot was born on Wednesday, September 26, 1888, but his twenty-eighth birthday was on *Tuesday*, September 26, 1916, because the intervening century year of 1900 was not a leap year. Lacking the century irregularity, however, the patterns continued.

FOR a breather, consider that the twenty-eight parrot, native to Australia, got its name because its call sounds like the number 28. Fortunately, its natural habitat is an English-speaking territory.

BACK to math: 28 is the second perfect number. It equals the sum of its proper factors: $(1 + 2 + 4 + 7 + 14 = 28)$.

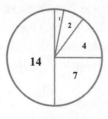

NOTE that the first perfect number, 6, equals $2(2^2 - 1)$, while $28 = 2^2(2^3 - 1)$. The ancient Greeks were aware that perfect numbers could be constructed in this fashion, but it took until 1849—in a posthumous paper of Leonhard Euler's—to reveal that *all* even perfect numbers must follow this same template: Namely, if p is a prime such that $2^p - 1$ is prime, then the number $P^n = 2^{(p-1)}(2^p - 1)$ is a perfect number. Primes of the form $2^p - 1$ are known as Mersenne primes, in honor of French monk Marin Mersenne (1588–1648). The first six Mersenne primes, together with the perfect numbers they generate, are listed below. Note that the perfect numbers get big awfully quickly, and it is not known whether the list is infinite. Nor has the existence of an *odd* perfect number ever been revealed or proved impossible. The hard part is the fact that any even perfect number has the form $2^{(p-1)}(2^p - 1)$ for some p. But that's just $\frac{2^p(2^p - 1)}{2}$, which is the sum of the first $2^p - 1$ positive integers and therefore triangular by definition.

n	P_n	P_n
1	2	6
2	3	28
3	5	496
4	7	8,128
5	13	33,550,336
6	17	8,589,869,056

▼

28 is also the fourth hexagonal number. The term doesn't quite mean what you'd think it would mean, as it refers to the combined number of dots in a bunch of nested hexagons. The concept does lend itself to a surprisingly easy formula, however, as the nth hexagonal number is given by the formula $H_n = n(2n - 1)$. It turns out that every hexagonal number is triangular (but not vice versa), with $H_n = T_{2n-1}$. Below, the first four hexagonal numbers are converted into the first, third, fifth, and seventh triangular numbers, with $28 = T_7$. (Note that the perfect numbers 6 and 28 are both triangular: While it might seem naïve to conjecture that all even perfect numbers must be triangular, that conjecture turns out to be true, and isn't even very hard to prove.)

▼

TO see the triangularity of 28 in a different context, a set of standard double-six dominoes has precisely 28 pieces: Each of the seven possibilities (0 through 6 and blank) is paired with each of the others, including itself, for a total of $7 + 6 + 5 + 4 + 3 + 2 + 1 = 28$ dominoes.

29 $\left[\text{ prime} \qquad (2 \times 9) + (2 + 9)\right]$

WHEN two primes differ by two, they are called twin primes, and 29 and 31 are the fifth such pair. The first five pairs of twin primes are (3, 5), (5, 7), (11, 13), (17, 19), and (29, 31).

Twin primes have held fascination to number theorists for many centuries, but surprisingly little is known about them. The most important open question is whether there are infinitely many such pairs.

One small step in that direction was advanced in 1919 by Viggo Brun, who demonstrated that twin primes aren't especially common within the world of primes. Brun's Theorem states that the infinite series $\frac{1}{3} + \frac{1}{5} + \frac{1}{7} + \frac{1}{11} + \frac{1}{13} \cdots$, where each term is the reciprocal of a twin prime, is a convergent series, meaning that the sequence of partial sums $\frac{1}{3}, \frac{1}{3} + \frac{1}{5}, \frac{1}{3} + \frac{1}{5} + \frac{1}{7} \ldots$, approaches a fixed limit. (That limit, called Brun's constant, is a number in the neighborhood of 1.902.)

Brun's result might seem singularly unimpressive for those who have never seen the harmonic series $\frac{1}{2} + \frac{1}{3} + \frac{1}{4} + \frac{1}{5} + \ldots$, formed by summing the reciprocals of every positive integer. The harmonic series *diverges*, meaning that the series of partial sums will exceed any fixed number if you go far enough out. In fact, even if you use only prime numbers, the sum $\frac{1}{2} + \frac{1}{3} + \frac{1}{5} + \frac{1}{7} \ldots$ is still unbounded. However, Brun showed that the partial sums of *twin* prime reciprocals converges—whether or not there are infinitely many of them!

Brun, being Norwegian, would have been very familiar with a 29-letter alphabet, made official in Norway in 1917 and now pretty much the standard throughout the rest of Scandinavia: the traditional 26-letter Latin alphabet plus three special vowels.

▼

IF surveyor A. P. Green had been in charge, the city of Twentynine Palms, California, might have had a different name. In an 1858 survey, Green found only 26 palm trees surrounding what is now the Oasis of Mara, three short of the unconfirmed count of 1855. Today, Twentynine Palms is best known as the home of the world's largest marine base as well as Joshua Tree National Park, immortalized by U2's landmark album, *The Joshua Tree*.

▼

THERE are 29 different pentacubes—objects created by joining five cubes at their faces. Whereas the 12 two-dimensional pentominoes (see **12**) can be arranged to create various rectangles, the 29 pentacubes cannot possibly form a three-dimensional block, or rectangular prism. Do you see why that's the case? (See Answers.)

▼

THE Lebombo bone is considered by many to be history's oldest mathematical artifact. The bone, excavated from the Lebombo Mountains in Swaziland during the 1970s, is a baboon tibia into which 29 notches were carved circa 30,000 BC. Although the precise utility of the bone is a matter of dispute, it resembles the counting or calendar sticks used in many primitive cultures. Some archaeologists have conjectured that the markings relate to the lunar or menstrual cycle of 29 days.

▼

FEBRUARY 29 is the leap day, and it occurs essentially once every four years. The concept of a leap year dates back to 45 BC, when Julius Caesar, acting on the advice of Alexandrian astronomer Sosigenes, mandated that the calendar adjust for the fact that the earth's journey around the sun takes 365.24 days. At the time, February was the last month of the year, so it made way for the extra day, though originally on the twenty-fourth of the month. February 29 became more and more entrenched as the centuries went by. According to a piece of Scottish folklore, that country's Parliament passed a law in 1288 establishing February 29 as the one time when a woman could propose marriage to a man. In Ireland, it has been said that St. Patrick refused a February 29 marriage offer from St. Bridget, although there is scant evidence for this claim. The modern leap-year standard was established in the 1500s by Pope Gregory XIII: In the Gregorian calendar, every year that is divisible by 4 is a leap year, except for century years such as 1900 that aren't divisible by 400 and—although it hasn't happened yet!—years divisible by 4000.

30 $\left[2 \times 3 \times 5 \right]$

THE stacking game Pylos, made by the French game company Gigamic, consists of 30 wooden balls: 15 light and 15 dark. In playing the game, the balls are placed on four levels, beginning with the 4×4 array at the bottom and ending with the solitary ball at the top. In mathematical terms, 30 is a square pyramidal number, the name for any number that can be expressed as the sum of the first n squares for some value of n—in this case $n = 4$.

▼

THERE are 360 degrees in a circle and 12 hours on a clock, and therefore $\frac{360}{12}$ = 30 degrees between any two adjacent hours. In three dimensions, however, things change. Because a complete rotation of planet Earth is 24 hours as opposed to a clock's 12, 30 degrees of longitude effectively represents *two* hours, not one.

▼

IF one of the angles in a right triangle is 30 degrees, the side opposite that angle is half as long as the hypotenuse.

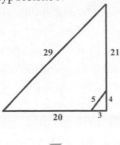

THERE are 30 distinct dice that can be created by arranging the numbers 1 through 6 on a cross-shaped figure and then folding along the lines to form a cube. The standard die, in which opposite faces sum to 7, is in gray, while the ninth cross reverses the position of 3 and 4.

Row 1:
2 / 3 1 4 / 5 / 6   2 / 3 1 5 / 4 / 6   2 / 3 1 6 / 4 / 5   2 / 3 1 6 / 5 / 4   2 / 3 1 4 / 6 / 5   2 / 3 1 5 / 6 / 4

Row 2:
2 / 4 1 5 / 3 / 6   2 / 4 1 6 / 3 / 5   2 / 4 1 3 / 5 / 6   2 / 4 1 6 / 5 / 3   2 / 4 1 3 / 6 / 5   2 / 4 1 5 / 6 / 3

Row 3:
2 / 5 1 4 / 3 / 6   2 / 5 1 6 / 3 / 4   2 / 5 1 3 / 4 / 6   2 / 5 1 6 / 4 / 3   2 / 5 1 3 / 6 / 4   2 / 5 1 4 / 6 / 3

Row 4:
2 / 6 1 4 / 3 / 5   2 / 6 1 5 / 3 / 4   2 / 6 1 3 / 4 / 5   2 / 6 1 5 / 4 / 3   2 / 6 1 3 / 5 / 4   2 / 6 1 4 / 5 / 3

Row 5:
3 / 4 1 6 / 5 / 2   3 / 4 1 5 / 6 / 2   3 / 5 1 6 / 4 / 2   3 / 5 1 4 / 6 / 2   3 / 6 1 5 / 4 / 2   3 / 6 1 4 / 5 / 2

ANY number that is less than 30 and relatively prime to 30 (sharing no common factors) must itself be prime. No number greater than 30 has this property—for example, 32 doesn't work, because 15 is less than 32 and shares no common factors, but 15 is not prime.

What makes 30 special is that it is the product of the first three primes. Note that the property doesn't extend to 210 (the product of the first four primes), because 210 exceeds products of primes such as 143 (11 × 13), which obviously have no factor in common with 210.

▼

AT one time a 30 signified the end of a wire service story. The precise reason for this use of 30 is unknown, although it has been traced to the Roman numeral XXX as well as to *fertig* (a German word meaning "finished" or "ready").

▼

30 is the sum of four consecutive numbers: 6, 7, 8 and 9. Not extraordinary perhaps, but to see this addition in action, check out the four sides of the Beatles' *White Album*, which consist of eight, nine, seven, and six songs respectively.

31 [prime]

AS we saw in **28**, a Mersenne prime is a prime of the form $2^p - 1$, with p prime. If $p = 5$, then $2^p - 1 = 31$, making 31 the third Mersenne prime, after 3 and 7.

Numbers of the form $2^n - 1$ crop up in solving the Tower of Hanoi puzzle, as depicted above. The puzzle was devised by famed mathematician Edouard Lucas in 1883, and the general form of the puzzle has three pegs and some number of discs sitting on one of the pegs, tapered so as to form a cone. The challenge is to move the discs one at a time to transfer the nested sequence onto another peg; the rub is that you cannot ever place a disc on top of a smaller one.

For the five-disc version shown here, a solution requires 31 moves. In general, for n discs, the minimum solution requires $2^n - 1$ moves. The more mathematically inclined will be interested to hear that solving the n-disc puzzle is equivalent to finding a Hamiltonian path on an n-hypercube.

THERE'S a different sort of puzzle, not quite as famous, that ends up highlighting the position of 31 as one less than a power of two. We start by choosing two points on the circumference of a circle and connecting them with a line segment. Obviously, that separates the circle into two parts. If we choose three points on the circle and connect every two of them, we separate the circle into four parts. If we continue, we get the following diagram, with the results summarized in the table below:

Number of points	Maximum number of pieces in circle
2	2
3	4
4	8
5	16

From the pattern thus far, choosing *six* points and connecting every possible pair with line segments seems destined to produce a total of 32 pieces, but in fact the maximum number of pieces is only 31. So close and yet so far. This unexpected result is only the beginning. Suppose the above situation arose in the context of a standardized exam, where you were asked to fill in the blank in the sequence 1, 2, 4, 8, 16, __ with one of the following choices:

A) 30   B) 31   C) 32   D) 33

Obviously if we choose anything other than C, we'll have some explaining to do. But *any* answer can work if you're creative enough. We have already seen that B can be justified by the points/lines/circles example. If you choose A, just tell the examiners that the sequence was the number of divisors of *n*!. (That's right: 1! = 1 has one divisor, 2! = 2 has two divisors, 3! = 6 has four, 4! = 24 has eight, 5! = 120 has sixteen, and 6! = 720 has thirty divisors.) And if you choose D, just say that you thought the sequence was the number of ways in which the first player gets killed in a five-player Russian roulette game using a gun having *n* chambers, where the number of bullets can equal anything from 1 to *n*, with no rotations of the cylinder allowed. That should give you extra credit.

▼

IF you'd rather see this same principle in a puzzle than something out of a standardized exam, here's one from the late, great puzzle inventor Nob Yoshigahara. With the understanding that you can't always trust patterns, all you have to do is figure out what number belongs in the circle with the question mark.

(See Answers.)

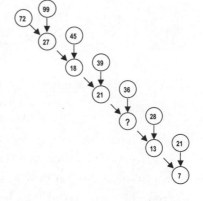

▼

WE saw in **4** that any positive integer can be represented as the sum of at most four perfect squares, not necessarily distinct. Well, there are only 31 numbers that cannot be represented as the sum of *distinct* squares. With the observation that 31 is one of these numbers, here is the entire list:

2 3 6 7 8 11 12 15 18 19 22 23 24 27 28 31 32 33 43 44 47 48 60 67 72 76 92 96 108 112 128

Now consider the following list, again 31 numbers long:

1 2 3 5 6 7 9 10 11 13 15 17 18 19 22 23 25 27 31 33 34 37 43 47 55 58 67 73 82 97 103

I came across this list online, alongside the claim that any positive integer could be written as the sum of four distinct squares . . . except for these numbers and any power of four multiplied by any of these numbers.

Now wait a minute. What was going on? A different set of 31 numbers, also linked to sums of distinct squares but with a different conclusion? That was a little weird. The key to understanding the claim is to observe that 14 is the first number that is neither on the list nor a power of 4 times a number on the list. And look what we have here: $14 = 9 + 4 + 1 + 0$. Evidently this second list governs what numbers cannot be represented as the sum of *precisely* four distinct squares . . . including zero!

32 $\left[\, 2^5 \,\right]$

I have to confess that when I starting writing this book, the number 32 was the sort of number I expected to write about at great length. After all, 32 is a power of 2, 32° Fahrenheit is the freezing point of water at sea level, 32 ft/sec^2 is gravitational acceleration in the non-metric world, and 32 is a popular uniform number for professional athletes, among them Magic Johnson, Sandy Koufax, and Jim Brown. But I don't have all that much!

▼

IN the world of sports and games, the number 32 isn't restricted to uniform numbers. The card game Skat uses a deck with no card lower than a 7, for 32 cards in all. In chess, there are 32 pieces on the board when the game begins, not to mention 32 white squares and 32 black squares. And, speaking of black and white, there are precisely 32 geometric shapes (12 pentagons and 20 hexagons) on a soccer ball. (See **12** for a more complete discussion of such shapes.)

▼

A compass rose indicates all 32 named directions—north, north by east, north-northeast, northeast by north, northeast, and so on clockwise around the circle. Because these various directions basically arise from bisecting a circle again and again, powers of two keep appearing, from the four basic directions N, E, S, and W all the way to the 32 directions represented.

33 $\left[\, 3 \times 11 \qquad 1! + 2! + 3! + 4! \,\right]$

THE UK version of peg solitaire is played on a board with 33 holes. A move consists of a jump, either horizontal or vertical, of one peg over another into an empty hole, at which point the jumped-over peg is removed. The objective is to continue in such a way that the final peg lands in the middle square.

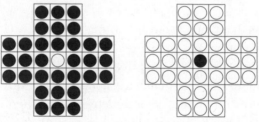

Because there are 32 pegs initially, a total of 31 jumps are required to go from the starting position to the winning position in the diagrams above. However, by combining jumps in a single move, as in draughts, only 18 moves are required to go from start to finish.

▼

IN the magic hexagram below, each arrow points to a set of five numbers adding up to 33.

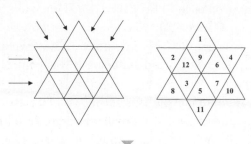

THE triangles in the picture above remind me that 33 is the largest number that cannot be represented as the sum of distinct triangular numbers. (After 33, we have 34 = 6 + 28, 35 = 1 + 6 + 28, 36 = 36, 37 = 1 + 36, 38 = 10 + 28, to name just a few.)

THE number 33 has found its way into cultures the world over. In Spanish, the phrase *"Diga treinta y tres"* ("Say 33") is used in the same way as is "Say cheese" in English. In Romania, doctors often ask their patients to say "33" (*"Treizeci si trei"*) when listening to their lungs with a stethoscope.

34 [2 × 17]

THE figure to the right was created by joining 34 dots in a special way. What makes the graph special is (1) it cannot be traced, meaning that it is impossible to draw a path that visits each dot once and only once (what mathematicians call a Hamiltonian path), and (2) if you remove any one dot,

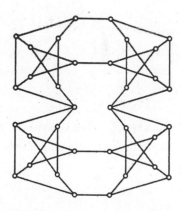

the resulting graph *can* be traced. Any graph with properties (1) and (2) is termed *hypotraceable*.

This construction on page 111 is called the Thomassen graph, after Danish mathematician Carsten Thomassen. Remarkably, it is the smallest known hypotraceable graph. (There do not exist any hypotraceable graphs with fewer than ten vertices, and for years it was conjectured that there weren't any, period.)

The theory of Hamiltonian paths is at the root of a classic problem called the Traveling Salesman Problem, in which the task is to find the cheapest route by which a salesman can make one stop in each of a set of cities and then return to the city in which he started. A more modern version of this problem arises in scheduling the route of a drill machine in manufacturing a printed circuit board. In robotic machining or drilling applications, the "cities" are parts to machine or holes (of different sizes) to drill, and the "cost of travel" includes time for retooling the robot (single machine job sequencing problem). As the number of holes (or cities) grows, the number of possible circuits becomes so vast that the computations cannot be handled by brute force, so a theoretical approach is required.

▼

THE most famous 4 × 4 magic square ever created is the one imbedded in Albrecht Durer's 1514 engraving, "Melancholia I." That square appears to the right.

Note that all rows, columns, and diagonals sum to 34. The number 34 is the magic constant for all 4 × 4 magic squares because the sum of the first 16 integers divided by 4 equals $\dfrac{(\frac{16(17)}{2})}{4} = 2 \times 17 = 34.$

16	3	2	13
5	10	11	8
9	6	7	12
4	15	14	1

Note also that the year of Durer's work is displayed in the middle of the bottom row of his magic square.

▼

A relative of the 4 × 4 magic square is the following simple piece of mathemagic. Select a number from the 16 numbered squares above. Say we start with 5. Now choose a number that is not in the same row or column as 5, say 15. Repeat this process once more, this time by choosing the number 2. (The third number cannot share a row or column with *either* of the first two choices.) There is one number

1	2	3	4
5	6	7	8
9	10	11	12
13	14	15	16

left that is not in the same row or column as any of the first three choices, and that is 12. Add 5, 15, 2, and 12 together and what do you get? Thirty-four, of course.

The reason the trick works is that the square can be constructed as the sum of a row and a column, as below. Every white square is obtained by adding the gray square to its left and the gray square above it. Therefore, if you have four white squares that share no row or column, their sum must be the sum of all the gray, which is $1 + 2 + 3 + 4 + 0 + 4 + 8 + 12 = 34$.

	1	2	3	4
0	1	2	3	4
4	5	6	7	8
8	9	10	11	12
12	13	14	15	16

EVERY two years voters across America elect members of the Senate. There are 100 senators in all, each serving terms of six years. Because the elections are divided up as equally as possible, the theoretical maximum number of Senate contests in any particular election year is 34.

This particular subdivision of 100 goes all the way back to Dante's *Divine Comedy* (see **3**). Of the 100 cantos in his masterpiece, 33 were devoted to the Sky, another 33 to Purgatory, and 34 to the most famous segment of all, Hell, otherwise known as Dante's *Inferno*.

THE number 34 is the ninth Fibonacci number (the sum of 13 and 21, the previous two members of the sequence). Field daisies are sometimes used to illustrate the appearance of Fibonacci numbers in nature (a phenomenon known as Ludwig's Law), as they often have 34 petals. Because daisies are used in the classic game of "she loves me, she loves me not," an even number of petals is undesirable. Fortunately, nature isn't always exact in its proportions, and in any event there are plenty of 13-petal and 21-petal flowers to go around.

35 $\left[\, 5 \times 7 \qquad (10-3)(10-5) \,\right]$

A figure made up of six squares joined at their edges is called a hexomino, and there are 35 of these altogether, assuming that each shape is considered equivalent to those obtained by rotating it or flipping it over. (See **11**.)

▼

IT is possible to place a knight on a standard 8 × 8 chessboard and then move the knight 35 times without crossing over its own path. Care to find that sequence? (See Answers.)

▼

IF you add the first five triangular numbers, you get 1 + 3 + 6 + 10 + 15 = 35. Geometrically, this means that if you place 15 billiard balls into their standard triangular rack, then place atop that triangle a triangle with four billiard balls on a side, and keep going, then the last of the 35 billiard balls will complete a triangular pyramid, or tetrahedron.

▼

LUDOLPH van Ceulen was a German mathematician who moved to the Netherlands and in 1600 became the first professor of mathematics at the

University of Leiden. When van Ceulen died in 1610, the number 3.14159265358979323846264338327950288, which you will recognize as π carried to 35 decimal places, was inscribed on his tombstone. Carrying out that calculation was van Ceulen's life's work.

Van Ceulen's approach was basically that of an obsessed Archimedes. The idea, in fact known to the Greeks of 200 BC, involved the inscription and circumscription of polygons in and around a circle. In English, we'll look at a simple case: the square.

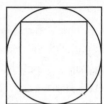

In the diagram to the right, we assume that the diameter of the circle is 1. The diagonal of the inscribed square therefore has length 1, so the side of that square has length $\frac{\sqrt{2}}{2}$, for a perimeter of $2\sqrt{2}$. The sides of the outer square are each equal to the diameter, or 1, so the perimeter there is 4. The point is that the circumference of the circle must lie somewhere in between those two perimeters. Because π, by definition, equals the circumference of a circle divided by its diameter, we get that $2\sqrt{2} < \pi < 4$, a valid inequality but, alas, only a crude approximation. Obviously you can improve the approximation by inscribing and circumscribing a regular polygon having more sides than a mere square, and that's precisely what van Ceulen did. His 35-digit approximation of π used polygons having 2^{62} sides! (No, he couldn't draw them. But he could calculate their perimeters, and that's all he needed.)

▼

THE Pont du Gard, near Remoulins in the south of France, is a Roman-style aqueduct with several rows of arches. Its topmost row consists of 35 small arches.

By happy but meaningless coincidence, there are 35 bridges across the Seine in Paris, beginning with the Pont est du Boulevard périphérique and ending with the Pont ouest du Boulevard périphérique, of course.

▼

THE number 35 also makes an appearance in the world of weights and measures. There are 35 Imperial gallons in a barrel of oil, and 35mm film has been an industry standard since the days of Thomas Edison and George Eastman.

36 $\left[\, 2^2 \times 3^2 \,\right]$

THE world-famous Rockettes consist of precisely 36 dancers. This number apparently derives from how many dancers can comfortably fit onstage to perform a kick line. (On the road, away from spacious Radio City Music Hall, the troupe is smaller.)

As it turns out, 36 dancers can be arranged either as a 6 × 6 square or as a triangle with a height of 8. The triangle formation was adopted in the Rockettes' performance of Happy Feet, not to be confused with the 2006 penguin movie of the same name. (The triangularity of 36 also arises within the rituals of Hanukkah, because by Hillel tradition one Menorah candle is lit the first night, two the second night, and so on, yielding a total of 1 + 2 + 3 + 4 + 5 + 6 + 7 + 8 = 36 separate lightings during the eight-day celebration.)

▼

36 is in fact the first number (not including the trivial 1) that is both square and triangular. The great Leonhard Euler demonstrated back in 1730 that there are infinitely many such numbers, but 36 is really the only one available to the Rockettes, as the next square triangular number is 1,225.

▼

SPEAKING of squares and triangles, the 6 × 6 square diagram on the next page displays the 36 possible outcomes of rolling a pair of dice. The gray circles contain the 21 distinct outcomes—not differentiating 2-5 (a 2 then a 5) from 5-2 (a 5 then a 2), and so on. The white circles contain the 15 repeated outcomes. By arranging the 36 possibilities in this fashion, we get a simple illustration of the fact that *any* perfect square can be written as the sum of two consecutive triangular numbers.

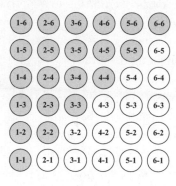

THE Lincoln Memorial in Washington, DC, has 36 columns—12 in the front and back, and 8 on each side. (Adding 12, 12, 8, and 8 together yields 40, not 36, but 4 must be subtracted from that total because each corner is counted twice.) The top of each column is inscribed with the name of one of the states of the union at the time of Lincoln's presidency; conveniently, the thirty-sixth state, Nevada, joined the union just days before Lincoln's reelection in 1864.

NOT far from the Lincoln Memorial, the Alexandria, Virginia-based game and puzzle company ThinkFun has found the number 36 just right on a couple of occasions. In 1996, they produced the 36 square-unit puzzle Rush Hour, devised by the inventor Nob Yoshigahara, whose diabolical number puzzle we encountered in **31**. The challenge in Rush Hour is to move surrounding cars and trucks so that

a red car can escape from a 6 × 6 grid. Then, in 2008, the company released 36 Cube, which challenges solvers to complete a solid cube out of 36 towers: six sets of towers, with each set consisting of six pieces of the same color but different heights. Labeled "The World's Most Challenging Puzzle," the median solving time has been estimated to be . . . never.

▼

THE 36 Officer Problem is an age-old conundrum that can be explained using this grid:

A	B	C	D	E	F
A	B	C	D	E	F
A	B	C	D	E	F
A	**B**	**C**	**D**	**E**	**F**
A	B	C	D	E	F
A	B	C	D	E	F

Note that we have placed six different versions of the letters A through F into a grid with 36 squares. The problem is to rearrange the letters so that no row or column contains two of the same letters or two of the same typeface. The grid below shows a solution for the 3 × 3 case. (The 2 × 2 case is easily seen to be unsolvable.)

A	C	B
C	**B**	A
B	A	**C**

The 36 Officer Problem gets its name because the problem was originally posed in terms of six regiments each containing six officers of different rank. Alas, the problem is unsolvable. That much was conjectured by Euler in 1780 or so, and was proved once and for all in 1901 by an amateur mathematician named Gaston Terry.

Euler had in fact speculated that so-called Graeco-Latin squares might be unsolvable for a wide range of square sizes, and it wasn't until 1959 that mathematicians discovered the astonishing truth that the *only* unsolvable squares are the trivial 2 × 2 case and the 6 × 6 case depicted on the previous page. That's right: A solution exists for any other square size whatsoever.

▼

THERE are 36 inches in a yard. As it happens, 36 inches is also the official height of the center of a tennis net (where the so-called center strap can be tightened or loosened to determine the overall height). In the days of wooden rackets, the height of a racket plus the width of the racket face came out to almost precisely 36 inches, and you'd often see players checking the net height that way. Once tennis rackets became oversized, however, that technique no longer worked. Today you seldom see the height of a net checked, even though having a simple yardstick on hand would serve that purpose.

▼

EACH angle of the five-pointed star below measures precisely 36 degrees. But you don't have to take my word for it.

Simply rearrange the five triangles as follows. Together, the tip angles sum to a half circle, or 180 degrees, so each one must measure $\frac{180}{5} = 36$ degrees.

▼

BASE 36, sometimes called the alphadecimal system, is a convenient positional numeral system because it uses each of the 26 letters of the alphabet coupled with the digits 0 through 9 (A = 10, B = 11, . . . Z = 35). Although base 36 is most useful in the context of computer programming (it is used by many URL redirection systems), it can also be used to give a unique number for any word or name. For example, GO = $16 \times 36 + 24 = 600$: G and O have positional values of 16 and 24, respectively, so the word GO corresponds to $16 \times 36 + 24 = 600$.

Unfortunately, because each place in the number represents a power of 36, the base 10 numbers associated with base 36 words are rather large: As unwieldy as NIEDERMAN is, for example, I'm not about to jettison it for 66,327,368,641,439. Because the idea of a positional numerical base is credited to the Babylonians (their preferred base was 60), I should point out that Babylon nemesis XERXES translates into 2020201444, but it's too bad that the Persian king of 20 years (485–465 BC) wasn't XERXUS instead, as his base 10 equivalent would then have been the far more satisfying 2020202020.

▼

NINETEENTH-CENTURY French writer George Polti wrote a book called *The 36 Dramatic Situations*, an attempt to classify every story line that could possibly arise in a play or book: "Crime pursued by vengeance," "Fatal imprudence," "Mistaken jealousy," and so on. Polti admitted that the choice of 36 was somewhat arbitrary, but he might have been modeling his list on "The Thirty-Six Strategies," a collection of military strategies from ancient China. The military strategies are surely more colorful, ranging from "(3) Kill with a borrowed knife" to "(15) Entice the tiger to leave its mountain lair" to perhaps the most famous of them all, "(36) If all else fails, retreat."

37 [prime]

$$37 = \frac{666}{(6 + 6 + 6)} \qquad 123{,}456{,}790 \times 3 = 370{,}370{,}370$$

THE curiosity that underlies these and other multiplication formulas involving the number 37 is that $3 \times 37 = 111$.

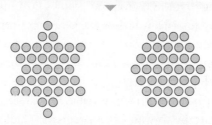

THE above arrangements show how 37 dots can be made into either a star or a hexagon. Note that a star of this type is constructed by first forming a hexagon, then adding six triangles of dots around the outside. Very few numbers can be represented as both a centered hexagon and a star—37 is the first number other than 1 to have this particular feature, and the next one doesn't come around until 1,261. The arrangement on the left is that of a French solitaire board, while the arrangement on the right corresponds to the dots on the mouthpieces of yesteryear's telephone handsets, or even the airholes of a bathroom vent on a Boeing 757.

IT is easy to see that 37 is the maximum number of points in a bridge hand. According to the standard point-count system, aces count as 4 points, kings 3, queens 2, and jacks count as one point. All told, the 16 honor cards yield a total of 40 high-card points for the whole deck, but a single hand can only accommodate 13 out of those 16 cards. Subtract three (one-point) jacks of your choice and you have your maximal 37-point hand—four aces, four kings, four queens, and a jack.

IF you've never seen the Sultan's Dowry Problem before, you're in for a treat. The setup is that a sultan has granted a commoner the chance to marry one of his hundred daughters. The commoner will be shown the daughters one at a time and will be told each daughter's dowry. The commoner has only one chance to accept or reject each daughter, meaning that he can't go back and choose one that he previously rejected. The catch is that the commoner may only marry the daughter who has the highest dowry of them all. Assuming that the commoner knows nothing in advance about the way the dowries are distributed, what is his optimal strategy to locate that special daughter?

Sounds kind of impossible, doesn't it? After all, the chance of any one daughter having the highest dowry is $\frac{1}{100}$, and nothing is known about the magnitudes and distributions of the dowries. But there is an optimal strategy, which turns out to be to wait until precisely 37 of the daughters and dowries have been shown, then to pick the next daughter whose dowry exceeds any of those seen thus far. Coincidentally, the chance of finding the highest dowry by following that strategy is about 37%.

▼

THE normal human body temperature, usually referenced as 98.6°F in the United States, is precisely 37° Celsius. At least, that's true if you follow the formula $F = (\frac{9}{5})C + 32$. But the original scientific work on human temperatures was conducted in the nineteenth century by the memorably named German physician Carl Reinhold August Wunderlich, who took the temperature readings of thousands of people for his statistical study. The figure 37°C is actually nothing more than Wunderlich's average temperature reading rounded off to the closest degree Celsius, meaning that our sacrosanct Fahrenheit equivalent of 98.6°F connotes far more accuracy than it is entitled to.

▼

WILLIAM Shakespeare wrote a total of 37 plays. Recall from our discussion in **36** that in the eyes of Georges Polti, there are precisely 36 dramatic situations. We can therefore state with 100% certainty what everyone already knows—that at least two of Shakespeare's plays must revolve around the same basic plot! Yes, I know—the multiple subplots yield even more repetition than that. But there's an important mathematical principle at work here, one that goes by the unassuming name of the pigeonhole principle (also known as the Dirichlet box principle). It basically states that if you have more than n objects (as in 37 plays) and you try to place them in n holes (as in 36 dramatic situations), at least one of the holes must have more than one object in it.

The great thing about the pigeonhole principle is that its applications run wide, even though the principle itself is completely obvious. Lest you think this is much ado about nothing, note that pigeonhole arguments can also be used to prove a wide variety of other results in the areas of geometry, elementary number theory, the game of bridge, and just plain common sense, including the following:

1. There are at least two people in Tokyo with the same number of hairs on their heads.
2. If you place five points inside an equilateral triangle that measures two inches on each side, you can always find a pair of points separated by no more than one inch.
3. If you choose any ten positive integers between 1 and 100, there will always be two disjoint subsets of those ten numbers (i.e., no elements in common) that have the same sum.

The pigeonhole questions got a little tougher, didn't they? Well, you are invited to take a crack at any or all of the above. (See Answers.)

38 [2 × 19]

THE number 38 can be written as the sum of two odd numbers in ten different ways, as on the next page:

$$
\begin{aligned}
38 \quad &= \quad 1 + \mathbf{37} \\
&= \quad \mathbf{3} + 35 \\
&= \quad \mathbf{5} + 33 \\
&= \quad \mathbf{7} + \mathbf{31} \\
&= \quad 9 + \mathbf{29} \\
&= \quad \mathbf{11} + 27 \\
&= \quad \mathbf{13} + 25 \\
&= \quad 15 + \mathbf{23} \\
&= \quad \mathbf{17} + 21 \\
&= \quad \mathbf{19} + \mathbf{19}
\end{aligned}
$$

Observe that each of the ten pairs of odd numbers contains at least one prime number (shown in bold). In other words, 38 cannot be written as the sum of two *composite* odd numbers. In and of itself, that may not seem so remarkable, but it turns out that 38 is the *largest* even number with that property.

The above assertion isn't all that hard to prove. We start by noting that the first few even numbers greater than 38 can be written as the sum of two odd composite numbers: 40 = 15 + 25, 42 = 15 + 27, 44 = 35 + 9, 46 = 25 + 21, 48 = 15 + 33. Obviously we can't keep going one even number at a time, but we don't have to. We can cover the numbers 50 through 58 simply by adding 10 to the first number in each of the above sums; for 60 through 68, we add 20, and so on. The reason this approach works is that any number ending in 5 (other than 5 itself) is composite. Admit it, that was easier than expected. Right?

By contrast, the assertion that any even number is the sum of two primes has proved extraordinarily difficult to prove. Prussian mathematician Christian Goldbach conjectured this result back in 1742 and it remains unsolved at this writing. At this point it is known that if an even number cannot be expressed as the sum of two primes it must be extraordinarily large, and it's fair to say that computers are working day and night in the search for a counterexample that no one other than the programmers themselves really wants to find.

▼

THE number 38 is also at the end of a very different type of list. When written in Roman numerals, 38 is expressed as XXXVIII. It just so happens that if you wrote down all possible Roman numerals in alphabetical order, XXXVIII would be the very last number you'd write.

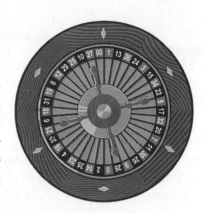

AN American roulette wheel has 38 slots, consisting of the numbers 1 through 36 plus 0 and 00. The primary effect of the extra zero slot is to increase the house advantage, as the probability of a gambler winning a red or black bet is $\frac{18}{38}$, smaller than the $\frac{18}{37}$ chance of winning such a bet on a European wheel. (See **37**.)

THE figure to the right is a magic hexagon, in which the numbers in the five columns, five left-slanting rows and five right-slanting rows all add up to 38.

Legend has it that this hexagon was constructed by retired railroad clerk Clifford Adams, who then passed it on to Martin Gardner at *Scientific American*. Gardner in turn

showed the construction to renowned recreational mathematician Charles Trigg, who confirmed that this magic hexagon is unique (any other solution of this size is a rotation/reflection of Adams's design).

Not only that, Trigg showed that the magic constant for a hexagon with n cells on each outside edge is given by the formula $\frac{9(n^4 - 2n^3 + 2n^2 - n)}{2(2n - 1)}$. Without getting carried away in the particulars, this expression can only be an integer if $\frac{5}{(2n - 1)}$ is an integer, which only happens for $n = 1$ and $n = 3$. In other words, Adams's creation is the only possible magic hexagon of any size, except for taking a single hexagon and sticking a "1" in it, and that's not so magical, is it?

39 $\left[\, 3 \times 13 \,\right]$

THERE are three different partitions of 39 that multiply to the same product, and 39 is the smallest number with this property. This is the Christmas Ribbon Problem with $n = 3$: Find three different package sizes with the same ribbon length and volume (see **118**):

$$39 = 4 + 15 + 20 \quad : \quad 4 \times 15 \times 20 = 1200$$
$$39 = 5 + 10 + 24 \quad : \quad 5 \times 10 \times 24 = 1200$$
$$39 = 6 + 8 + 25 \quad : \quad 6 \times 8 \times 25 = 1200$$

If you list all the primes between the smallest prime factor of 39 and the largest prime factor of 39, you get 3, 5, 7, 11, and 13. And 3 + 5 + 7 + 11 + 13 = 39. Neat trick, and it doesn't happen again until 155. (There's actually a number smaller than 39 with this same property. Can you find it?) (See Answers.)

▼

THE 1935 Alfred Hitchcock film *The 39 Steps* was based on a book by the same name by John Buchan. In the movie, the title refers to the name of a spy organization, while the book's title refers to a coastal site in Kent, where a path from a cliff to the water has precisely 39 steps. Along this latter line, perhaps it is worth noting that at the old Wembley Stadium, a winner had to ascend 39 steps in order to reach the Royal Box and receive the appropriate trophy.

▼

THE 39 Articles of Religion were established by the Anglican church in 1563 and even today form the basis for the Anglican faith. Somewhere in between, the United States of America declared its independence and wrote its own Constitution, which was ultimately signed by 39 men.

▼

39 is the highest number on a standard Master combination lock.

▼

A bowling lane consists of precisely 39 strips, usually made of wood, each measuring slightly more than an inch, for 42 inches altogether.

▼

IN bridge, there are only 39 possible distributions, or hand patterns. The most common distribution is 4–4–3–2, meaning that the hand consists of four of one suit, four of another, three of a third suit, and two cards in the fourth suit.

40 $\left[\, 2^3 \times 5 \,\right]$

RELIGION is replete with references to the number 40, from the 40 days of Lent to the traditional 40-day period of mourning in the Muslim faith. But the number 40 in the Bible sometimes refers to just a really, really big number, as opposed to a specific quantity, as in the 40 days and 40 nights of the Great Flood. Similarly, the use of 40 in the story "Ali Baba and the Forty Thieves" was less than literal, and the expression "40 winks" just meant a lot of sleep. It is a measure of societal inflation that the 40 of olden days has come to be today's "gazillion."

▼

ONE surprisingly literal use of the number 40 is found in the word *quarantine*. The word looks suspiciously like *quarante*, the French word for "forty" and the original Roman *quarantine* is said to have kept ships isolated in the harbor for forty days.

THE expression "Life begins at forty" is an outgrowth of a 1932 book by Walter Pitkin and a 1937 song by Sophie Tucker. The expression certainly applied to Princeton mathematician Andrew Wiles, a native Brit who had just turned 40 years of age when he gave his famous "Modular Forms, Elliptic Curves, and Galois Representations" lecture at Cambridge University in 1993: This was the lecture in which Wiles proved Fermat's Last Theorem—the assertion that the equation $x^n + y^n = z^n$ has no solutions in positive integers for $n > 2$ (see **2**). Unfortunately for Wiles, the Fields medal, perhaps the most prestigious award in higher mathematics (and the one credited to Matt Damon's mentor in *Good Will Hunting*), is only given out to people under 40, and when his original proof was revealed to have a gap, Wiles was essentially out of the running for the Fields class of 1994 (the medal is bestowed every four years). However, the International Mathematical Union gave Wiles a special plaque concurrent with the Fields Medals of 1998, by which time the gap in Wiles's proof, and his place in mathematical history, had long been resolved.

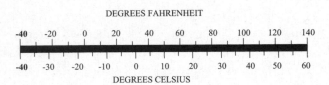

AS the drawing indicates, -40° Celsius is the same as -40° Fahrenheit, the only time the two temperature scales coincide. In general, the two scales are connected via the formula $F = (\frac{9}{5})C + 32$.

IN 1953, three scientists at the start-up Rocket Chemical Company were working on a compound to eliminate rust and corrosion on rockets and other metal parts, using a technique called water displacement. On their fortieth try, they succeeded and in so doing created their first commercial product. The company was renamed in 1969 to honor their flagship (and, at the time, only) product. The WD-40 Company finally expanded its product line through a series of acquisitions beginning in 1995. Within a few years, its offerings included such familiar consumer brands as 3-IN-ONE oil, Lava soap, 2000 Flushes, and Carpet Fresh.

▼

A B C D E F G H I J K Ⓞ M N O P Q Ⓡ S T U V W X Y Z

40 is the only number whose letters are in alphabetical order when written out in English.

41 [prime]

THE expression $x^2 - x + 41$ looks innocuous enough, but it has a remarkable property first noticed by Euler. Try plugging in some numbers for x, starting with 1, and see what the output looks like:

x	$x^2 - x + 41$
1	41
2	43
3	47
4	53
5	61
6	71
7	83
8	97
9	113
10	131

The first two numbers on the list, 41 and 43, form a set of twin primes—primes that differ by just two. Then you will note that the differences between successive terms go up by two with each step; $47 - 43 = 4$, $53 - 47 = 6$, and so on. But what makes this sequence remarkable is that *all* the numbers thus far are prime. In fact, if you kept going, you'd get a streak of *40* consecutive primes, ending with $40^2 + 40 + 41 = 1681$. (Note that you obviously get a composite result when $x = 41$, because $41^2 - 41 + 41$ equals 41^2.) While there are polynomial expressions that yield more than 41 consecutive primes, Christian Goldbach proved in 1752 that that no polynomial with integer coefficients can possibly yield a prime for all integer inputs x.

▼

A related sort of structure, known as a prime spiral, was apparently discovered by Polish mathematician Stanislaw Ulam while doodling during a boring meeting. His creation began by placing a 1 in the center of a grid and continuing outward with consecutive integers. He noticed that primes in such spirals often create interesting patterns. Had he started with the number 41 in the center, he would have come up with the following diagram, in which primes are shaded. Continuing this spiral would have produced a 40 × 40 square in which every element along this diagonal was prime—the very same primes doled out by Euler's quadratic!

45	44	43	
46	41	42	
47	48	49	50

▼

THE pattern to the right consists of 25 white squares and 16 black squares. A similar pattern can be created for any number that, like 41, is the sum of consecutive squares. But one nicety unique to 41 is that $41 = 4^2 + 5^2 = 1^2 + 2^2 + 6^2$, and the rightmost equality remains true even if the exponents are removed.

42 $\left[\, 2 \times 3 \times 7 \,\right]$

THERE are 42 eyes in a deck of cards: three two-eyed kings, one one-eyed king, four (all two-eyed) queens, two two-eyed jacks, and two one-eyed jacks makes a total of 21, and this number must be doubled because the images on the face cards appear twice on each card. (Speaking of games and doubling, we have seen that there are a total of 21 dots on a die, and therefore 42 dots on a pair of dice.)

▼

CRICKET, whose regulations are entrusted to London's Marylebone Cricket Club, has 42 official laws, with Law 42 governing fair and unfair play. In America, 42 was the number worn by Jackie Robinson during his career with the Brooklyn Dodgers. Robinson's number was retired by the Dodgers in 1972, just months before he died at the age of 53, and it was retired by all of Major League Baseball in 1997, with a grandfather clause exempting those who were already wearing the number at that time. Longtime Yankee closer Mariano Rivera has continued to wear number 42 for over a decade. Another longtime relief pitcher—Bruce Sutter of the Cubs, Cardinals, and Braves—also wore number 42 and was the only other player to have it retired.

▼

IN Douglas Adams's *Hitchhiker's Guide to the Galaxy*, the computer Deep Thought was asked to calculate the Ultimate Answer to the Great Question of Life, the Universe and Everything, and the answer was "42." The following memorable line was thereby uttered:

> "I checked it very thoroughly," said the computer, "and that quite definitely is the answer. I think the problem, to be quite honest with you, is that you've never actually known what the question is."

The Observer once called Adams "the Lewis Carroll of the twentieth century," perhaps unaware that the number 42 held special fascination for Carroll as well. The original *Alice in Wonderland* had 42 illustrations, the famous "All persons more than a mile high to leave the court" was Rule 42, and the Baker in "The Hunting of the Snark" had 42 boxes.

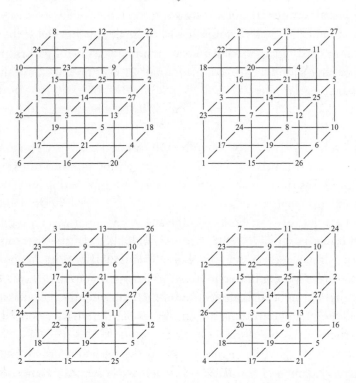

THE constructions above represent the four possible 3 × 3 magic cubes. (Technically they are called semi-perfect magic cubes or Andrews cubes, after early twentieth-century cube pioneer W. S. Andrews. All other 3×3 solutions are rotations or reflections of these four.) In each case the sum of the three numbers in any row or column equals 42, as does the sum of any three numbers occupying the same square in the three different sheets, as does the sum of any three numbers along a diagonal of the cube beginning and ending with one of the corners—but not the diagonal of any face. The magic constant of the magic cube is 42 because 42 equals the sum of the first 27 integers divided by 9. In general, the magic constant for an $n \times n$ magic cube equals $\frac{n(n^3 + 1)}{2}$. Note that $42 \div 3 = 14$ is the center number for all of the cubes.

The structure of the magic cube somehow reminds me that there were 42 prison cells in the infamous D Block at Alcatraz Prison, comprising 36 segregation cells and 6 solitary confinement cells. The D Block was the prison's maximum security area and was reserved for the worst offenders of the prison, which is saying a lot. The one redeeming feature of the cells was that they were bigger than found elsewhere in the prison. Block 42 was the longtime home of Robert Stroud, aka the Birdman of Alcatraz, who developed his interest in birds at Leavenworth Prison before entering the less ornithologically friendly confines of Alcatraz in—you guessed it—1942.

43 [prime]

BACK in the good old days, when Chicken McNuggets were available only in lots of 6, 9, 20, you could have asked the question "What is the largest number of McNuggets you *can't* buy?" The answer was 43. This type of problem has actually been around for some time, enabling mathematicians to call 43 the *Frobenius* number of the set {6, 9, 20}, thereby honoring German mathematician Ferdinand Georg Frobenius (1849–1917).

Strictly speaking, what's being sought is the highest number that isn't a linear combination of 6, 9, and 20—that is, cannot be expressed in the form $6a + 9b + 20c$, with a, b, and c all positive integers. The question makes sense only if the three numbers, as here, don't have a common factor. By way of contrast, if you combine a nickel, dime, and quarter, you'll always get a multiple of 5 cents.

Unfortunately, the introduction of a Happy Meal box of four Chicken McNuggets ruined everything. Once you could buy a box of four separately, the maximum number of McNuggets that can't be bought (using combinations of 4, 6, 9, and 20) is 11.

▼

WHICH reminds me: If you read our discussion in **11,** you know that there is an easy expression for Frobenius numbers for *two* variables: namely, the largest number that cannot be created using combinations of x and y equals $xy - x - y$. It turns out that having three (or more) numbers makes the problem substantially harder, and finding the Frobenius number is accomplished by use of an algorithm rather than a single formula, though there are formulas that work if the initial set of numbers is friendly enough. The problem is in fact difficult enough that certain classes of algorithms will always be inadequate because of the computer time required to implement them.

The general Frobenius Problem has applications far outside Chicken McNuggets. In economics, the question might be to gauge the possible outputs of a set of production or cost functions that take on integer values. In mass spectrometry (a technique to identify and/or quantify molecular compounds), the question might be to compute what types of molecular configurations might give rise to so-called peaks found in the data. All of which makes for interesting gatherings at Frobenius conferences.

▼

THERE'S one other appearance of 43 that's more of a curiosity than anything else. Recall that the Fibonacci sequence starts out 1, 1, 2, 3, 5 . . . with each successive term being the sum of the two that preceded it. From the same start, let's define a new sequence by making the sixth term equal to the sum of the squares of the first five terms, divided by the number $(5-1)$. We get $\frac{(1^2 + 1^2 + 2^2 + 3^2 + 5^2)}{4}$, which equals 10. (Note that the 2, 3, and 5 in the sequence are in fact generated by this same formula!) The seventh term equals $\frac{(1^2 + 1^2 + 2^2 + 3^2 + 5^2 + 10^2)}{5} = 28$, and so on. The sequence starts to grow quite rapidly starting with the tenth term. But because the creation of the sequence involves division, there's no reason in the world why the numbers produced should be integers. In fact, though, the first 43 members of the sequence are integers. I'd show you the first non-integral result, but it's too big to fit on this page—even if I had started at the top.

44 $\left[\, 2^2 \times 11 \,\right]$

AN Euler brick is a rectangular block in which the three sides a, b, and c are all integers, and the resulting diagonals d_{ab}, d_{ac}, and d_{bc} are also all integers. The smallest possible Euler brick has sides of length 44, 117, and 240, with diagonals of 125, 244, and 267. This

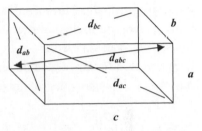

brick was discovered in 1719 by German mathematician Paul Halcke, but somehow Euler got his name on the whole concept even though he was only 12 years old at the time.

The diagram above actually has seven labeled lines, not six, the seventh being the "space diagonal" (with arrows) d_{abc}. Alas, the space diagonal in this diagram is the square root of 73,225, which is not a whole number. Remarkably, despite the long and storied history of the Euler Brick Problem, even today it is not known whether there exists *any* Euler brick whose space diagonal is an integer.

▼

ON Bastille Day of 1951, Frenchman A. Ferrier proudly announced that the 44-digit number $\frac{2^{148} + 1}{17}$ = 20,988,936,657,440,586,486,151,264,256, 610,222,593,863,921 was prime. Ferrier thus broke the 75-year record held by Edouard Lucas, who had checked the primality of $2^{127}-1$ by hand. Ferrier confirmed his number's primality by using only a desk calculator, and his discovery has the distinction of being the largest-known prime number calculated without any type of electronic computing device. Within a month of his announcement, a new age had dawned, and a 79-digit number was discovered by computer. As I began this book, the largest known prime (coincidentally, the forty-fourth Mersenne prime; see **28**) had 9,808,358 digits.

▼

IF you have five letters and five pre-addressed envelopes, in how many ways can you place the letters into the envelopes so that none of the letters is in the correct envelope? This problem was posed and solved by Pierre de Montmart in the early eighteenth century. De Montmart was a colleague of Nicholas Bernoulli, who solved the same problem using a set-counting formula called the inclusion-exclusion principle. De Montmart also knew Blaise Pascal and is credited with giving Pascal's Triangle its name. Permutations of this sort are called derangements, and the general formula for the number of derangements of *n* objects is given by the following formula:

$$n! \sum_{k=0}^{n} \frac{(-1)^k}{k!}$$

In particular, if $n = 5$, $n! = 120$, and the formula yields $120(1 - 1 + \frac{1}{2} - \frac{1}{6} + \frac{1}{24} - \frac{1}{120}) = 44$.

45 $\left[3^2 \times 5 \right]$

THE number 45 is triangular, being the sum of the numbers 1 through 9. In particular, 45 is the sum of the numbers in any row or column of a finished Sudoku grid. And, speaking of triangles, in an isosceles right triangle (where two legs of

equal length surround the 90-degree angle), both acute angles have a measure of 45 degrees.

▼

ANY triangular number is, by definition, the sum of the first *n* integers for some *n*. What makes 45 special is that it is the first number that can be represented as the sum of consecutive positive integers in six different ways. The representations are 45, 22 + 23, 14 + 15 + 16, 7 + 8 + 9 + 10 + 11, 5 + 6 + 7 + 8 + 9 + 10, and, finally, 1 + 2 + 3 + 4 + 5 + 6 + 7 + 8 + 9. It turns out that the number of representations of a number as the sum of consecutive integers is the same as the number of odd divisors of that number. For 45, we can enumerate the six expected divisors: 1, 3, 5, 9, 15, and 45 itself.

▼

ONE final item on the subject of numbers adding up to 45. Check out the square of numbers to the right. It is a magic square because every row, column, and diagonal adds up to the same number—45.

5	22	18
28	15	2
12	8	25

Of course, by itself that's nothing special. It's easy to generate magic squares with a constant of 45 because any genuine 3 × 3 magic square (using the numbers 1 through 9) will add up to 15 (see **15**), so you can just multiply every number by 3 to get 45. Of course, it's easy to see that the above square isn't of that sort. Actually, when you look further, this square doesn't look all that special in terms of the variety of the numbers, because every one of them ends in 2, 5, or 8. But check this out. Starting with the upper left, the number of letters in *five* is 4. The number of letters in *twenty-two* is 9. Look at the new square we create by continuing in this fashion:

4	9	8
11	7	3
6	5	10

The new square is itself a magic square, with magic constant 21. Words fail me.

▼

A 45-degree angle plays a theoretical role in the shot put and hammer throw events in track and field. In theory, a 45-degree landing angle is associated with the greatest distance for any release velocity. In real life, however, this figure is slightly reduced, for the simple reason that the objects are released several feet above ground level. And for events such as the discus and javelin, where aerodynamics play a more significant role, the 45-degree figure is all the more elusive.

THE forty-fifth parallel is halfway between the North Pole and the equator, although the flattening of the Earth near the poles creates a slight error.

THE number 45 can be broken into two parts that form its square, three parts that form its cube, and four parts that form its fourth power. No other number less than 400,000 has this property. (Numbers that have this property for any particular power n are known as Kaprekar numbers of order n.)

$$(20 + 25)^2 = 2025$$
$$(9 + 11 + 25)^3 = 91125$$
$$(4 + 10 + 06 + 25)^4 = 4100625$$

46 $\left[\, 2 \times 23 \,\right]$

IN New York State, a "46er" is someone who has climbed all 46 of the Adirondack mountains. The peaks range from Mount Couchsachraga, at 3,820 feet, to Mount Marcy at 5,344 feet.

THERE are 46 ways of using all 9 nonzero digits to create a fraction that equals precisely $\frac{1}{8}$. No other fraction of the form $\frac{1}{n}$ comes close to such a tally. In fact, these pandigital constructions for $\frac{1}{n}$ are impossible for all but a finite number of choices for n. (See **68**.)

▼

1. God is our refuge and strength, a very present help in trouble.
2. Therefore will not we fear, though the earth be removed, and though the mountains be carried into the midst of the sea;
3. Though the waters thereof roar and be troubled, though the mountains **shake** with the swelling thereof. Selah.
4. There is a river, the streams whereof shall make glad the city of God, the holy place of the tabernacles of the most High.
5. God is in the midst of her; she shall not be moved: God shall help her, and that right early.
6. The heathen raged, the kingdoms were moved: he uttered his voice, the earth melted.
7. The LORD of hosts is with us; the God of Jacob is our refuge. Selah.
8. Come, behold the works of the LORD, what desolations he hath made in the earth.
9. He maketh wars to cease unto the end of the earth; he breaketh the bow, and cutteth the **spear** in sunder; he burneth the chariot in the fire.
10. Be still, and know that I am God: I will be exalted among the heathen, I will be exalted in the earth.
11. The LORD of hosts is with us; the God of Jacob is our refuge. Selah.

Above is Psalm 46 from the King James Bible. In case you're wondering what it's doing here, the answer is that exhibits a Bible Code of a different sort from the one you may have read about. In Psalm 46, the forty-sixth word from the beginning is *shake* and the forty-sixth word from the end is *spear*. And in the year when the King James Bible came out—1610— William Shakespeare celebrated his forty-sixth birthday.

The above is nothing more than coincidence, but there is one age-old story involving 46 that happens to be true. Namely, 46 BC was the longest year in history, because that's when Julius Caesar adopted the Julian Calendar and in so doing created a year with 445 days.

▼

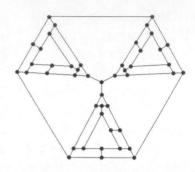

THIS diagram may look like a spiderweb, but it's actually the Tutte graph, a famous counterexample devised in 1946 by William Thomas Tutte, then a graduate student in mathematics at Cambridge University. By that time in his life, Tutte had already broken a vital German code system, altering for the better the course of the Allied invasion of Europe. The Tutte graph is a more humble sort of achievement. It is a graph with 46 vertices, each of which is connected to two other vertices, and with the property that any two vertices can be removed and still leave a connected graph (which means what you think it means). Peter Tait of the University of Edinburgh had conjectured in 1880 that any graph with these two properties must be Hamiltonian—that is, starting with any vertex, you can devise a path that takes you through each and every other vertex precisely once before returning to your starting point. The Tutte graph, which is not Hamiltonian, showed that Tait's conjecture was false. Too bad, because, in a great example of how different-looking pieces of mathematics are in fact intimately related, Tait's conjecture, if true, would have implied the Four-Color Map Theorem! (See **4**.)

47 [prime]

A cube cannot be divided into 47 subcubes, and 47 is the largest number with that property. This funny-looking result was proved in 1977, resolving a 30-year-old problem that went under the name of the Hadwiger Problem, after Swiss mathematician Hugo Hadwiger (1908–1981).

Part of the proof isn't all that hard. It begins with the curious observation that it is possible to divide a cube into 1, 8, 20, 38, 49, 51, or even 54 subcubes, based on the following equations:

$$1^3 = 1^3 \qquad\qquad\qquad 1 = 1$$
$$2^3 = 8 \times 1^3 \qquad\qquad\qquad 8 = 8$$
$$3^3 = 2^3 + 19 \times 1^3 \qquad\qquad 1 + 19 = 20$$
$$4^3 = 3^3 + 37 \times 1^3 \qquad\qquad 1 + 37 = 38$$
$$6^3 = 4 \times 3^3 + 9 \times 2^3 + 36 \times 1^3 \qquad 4 + 9 + 36 = 49$$
$$6^3 = 5 \times 3^3 + 5 \times 2^3 + 41 \times 1^3 \qquad 5 + 5 + 41 = 51$$
$$8^3 = 6 \times 4^3 + 2 \times 3^3 + 4 \times 2^3 + 42 \times 1^3 \qquad 6 + 2 + 4 + 42 = 54$$

The next step, equally curious, is to note that if the numbers m and n have the property that a cube can be divided into m and n subcubes, then the number $m + n - 1$ must have the same property. Just divide the original cube into m subcubes, then divide one of the subcubes so obtained into n subcubes, for a total of $m + n - 1$ subcubes. (Strangely, it is never possible to cut a cube into subcubes that are all of different sizes—no matter how many subcubes are involved, at least two of them must be identical.)

What makes the initial set of numbers special is that by starting with {1, 8, 20, 38, 49, 51, 54} and applying the formula $m + n - 1$ successively, it turns out to be possible to create *any* number greater than 47—for example, $57 = 20 + 38 - 1$, and so on. I'll leave the proof of that fact to the reader, and likewise for the demonstration that 47 subcubes *can't* be attained. (Remember, all we showed is that any number > 47 *can* be attained.)

▼

IF you'd prefer an easier challenge, 47 smaller triangles are nested in this triangle. Identifying the 47 is painstaking but not that difficult. Tougher would be to count the strings on a concert pedal harp, though I will spare you the challenge by revealing that there are 47.

IN 1964, Pomona College professor Donald Bentley "proved" that all numbers are equal to 47, the beginning of a long association between the college and that particular number. Here is a sampling of some 47 trivia accumulated on the Pomona website:

Pomona College is located off the San Bernardino Freeway. The sign reads: Claremont Colleges Next Right, Exit 47.

Claremont McKenna College, named after Donald McKenna '29, was founded in 1947.

There are 47 pipes in the top row of the Lyman Hall organ.

The Declaration of Independence consists of 47 sentences.

The Disney comedy *The Absent-Minded Professor* features a basketball game filmed at Pomona's old Renwick Gym. The final score: 47–46.

Harwood Dormitory has rooms 45 and 49, but (mysteriously) no Room 47.

In the film *The Towering Inferno*, actor Richard Chamberlain '56 was the 47th person in line to be rescued.

The Pythagorean Theorem is Proposition 47 of Euclid's Elements.

There are 47 letters in the dedication plaque on Mudd-Blaisdell Hall, which was completed in 1947.

At the time of Pomona's first graduating class (in 1894), there were 47 students enrolled.

If all this 47 trivia upsets your stomach, you'll be glad to know that Rolaids absorbs 47 times its weight in excess acid.

At risk of bursting Pomona's bubble, it is worth pointing out that pretty much any number is capable of producing a string of coincidences such as those above. But one area where the appearance of 47 is not a coincidence is in *Star Trek: The Next Generation*, where the crew stops at Sub-Space Relay Station 47, Data is unconscious for 47 seconds, a main character shrinks to 47 centimeters, there's a planet of 47 survivors, the crew discovers element 247, and many more. These selected mentions have the same feel as those on the Pomona website, but apparently they were instigated by Joe Menosky, a writer/coproducer of *Star Trek* whose handiwork has been continued by subsequent production teams. Menosky is a 1979 graduate of Pomona College.

48 [$2^4 \times 3$]

48 is the smallest number with ten divisors: 1, 2, 3, 4, 6, 8, 12, 16, 24, and 48 itself. (In general, $2^{n-1} \times 3$ is the smallest number with $2n$ divisors, as long as $n \geq 2$.)

48 is twice the total number of major and minor keys in Western tonal music (twenty-four), not counting enharmonic equivalents. Johann Sebastian Bach's *Well-Tempered Clavier* is informally known as *The Forty-Eight* because it consists of a prelude and a fugue in each major and minor key, for a total of 48 pieces.

WITH the addition of Arizona and New Mexico in 1912, the United States comprised 48 states, and the stars on the flag could be arranged in six rows of eight apiece. This configuration was maintained until Alaska joined the Union in 1959. Even today the continental United States is referred to as the "lower 48."

THIS rectangle contains 48 square units, half of which are gray and half of which are white. There is only one other rectangle that can be lined in this same fashion using an equal number of two different colors. Can you find it? (See Answers.)

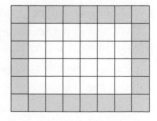

49 $\left[\ 7^2 = (11 - 4)^{(11 - 9)}\ \right]$

IT is possible to arrange 49 triangles (including the upside-down white ones) to form one big triangle of the same shape. But this construction can be done for *any* square number, not just 49. The principle here is that any perfect square can be written as the sum of two consecutive triangular numbers, as we saw in **36**. In this case 49 = 28 (gray) + 21 (white).

Alternatively, if you count the number of triangles (both white and gray) in each row, as you progress downward you get 49 = 1 + 3 + 5 + 7 + 9 + 11 + 13. In fact, *any* square can be written as the sum of consecutive odd numbers starting with 1.

But 49 is a very special square, as it is formed by joining two squares, 4 and 9, whose product is itself the perfect square 36.

▼

NOT only is 49 a square, it also has the peculiar property that if you put 48 in the middle you get another square, 4489. Repeating the process yields 444889, another square, and 44448889, yet another square.

▼

IN the standard game of lotto, six balls are drawn from a set that is numbered from 1 to 49. Given that the balls are not returned to the drum (in the probability world this is called "sampling without replacement"), the total number of choices equals $\frac{49!}{6!43!}$ = 13,983,816. This quantity is written as either $_{49}C_6$ or, using the more old-fashioned notation $\binom{49}{6}$, pronounced, descriptively, "49 choose 6." In general, "*n* choose *k*" equals $\frac{n!}{k!(n-k)!}$ and means just what you'd think: the number of ways of choosing *k* objects from an original set of *n*.

▼

THERE were two gold rushes in the United States during the nineteenth century. The first of these—the California Gold Rush of 1849—led to the term *49er* to describe a miner, immortalized in the song "Oh My Darling, Clementine" and by the San Francisco football team of the same name. In the last few years of the nineteenth century, the gold rush turned to the Klondike River region of Alaska, which eventually became the forty-ninth state.

50 $\left[\begin{array}{cc} 2 \times 5^2 & (10-5)(10-0) \end{array} \right]$

THE title of the hit TV show *Hawaii Five-O* derives from Hawaii being the fiftieth state. (Obvious, I know, but I have to confess that I never thought of that during the entire run of the series.) The US flag adopted its current configuration to accommodate the addition of Hawaii. The grid of 50 stars is essentially a 4 × 5 array nested inside a 5 × 6 array.

▼

THERE have been a total of 27 different star configurations on US flags, ranging from the original 13 stars to today's 50 stars. There was actually a 49-star flag that followed the flag in **48**; although Alaska and Hawaii were both admitted to the Union in the same year (1959), their admission dates straddled July 4, and, according to an Act signed by President James Monroe in 1818, the flag officially gets updated on the July 4 following the admission of a new state. The 50-star version therefore became official on July 4, 1960, and on July 4, 2007, it became the longest-lived flag in US history.

▼

THE book of Genesis contains 50 chapters.

▼

IN darts, the bull's-eye (innermost circle known as the bull) is worth 50 points.

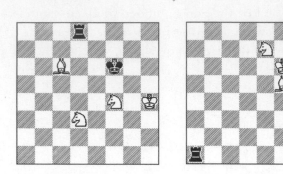

IN chess, a player can claim a draw if no piece has been captured and no pawn has been moved for the previous 50 moves. This rule was apparently first proposed in 1561 by Ruy Lopez, whose name now adorns a classic chess opening. Some 430 years later, Anatoly Karpov and Garry Kasparov battled for over 50 moves after reaching an end position in which Karpov held two knights and a bishop while Kasparov held just a rook (left). What made the game's conclusion especially fascinating is that Kasparov, perhaps unaware that the 50-Move Rule applied (a draw is not automatically granted—a player has to claim it), offered his rook as a sacrifice (two moves after the position on the right). But a draw was reached anyway. Had Karpov taken the rook, a stalemate would have resulted, while turning down the sacrifice would have reduced the end game to just two knights for Karpov, and it is impossible for a king and two knights to force checkmate!

IT is well-known that $50 = 1^2 + 7^2 = 5^2 + 5^2$ and is the smallest number that can be expressed as the sum of two squares in two different ways. But there's a geometric interpretation to this fact that isn't quite as familiar. Check out the following diagrams:

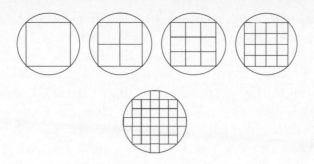

The top-left diagram shows a square inscribed within a circle. Obviously you can't insert any more squares of the same size. Ditto for the entire top row, ending with a 4 × 4 square. But when you circumscribe a 5 × 5 square with a circle, four more small squares fit inside the circle, as shown by the diagram in the bottom row. But that's just another way of saying that the hypotenuse of a right triangle with sides 1 and 7 is the same as the hypotenuse of the isosceles right triangle with sides 5 and 5, and that's what the equation $50 = 1^2 + 7^2 = 5^2 + 5^2$ is all about.

51 [3 × 17]

THE number 1,979,339,339 is a prime number with the rare property that if you take any number of digits off the right-hand side, you're still left with a prime . . . as long as you consider 1 to be a prime. Such numbers are called right-truncatable primes or, more colorfully, Russian doll primes, presumably in recognition of those little nested wooden dolls that can be taken apart repeatedly only to reveal another, smaller doll inside.

Of course, you may be wondering what all this has to do with the number 51. The answer is that there are precisely 51 Russian doll primes, with 1,979,339,339 being the largest of the lot.

▼

A somewhat better-known group of 51 is the following set of countries: Argentina, Australia, Belgium, Bolivia, Brazil, Byelorussia, Canada, Chile, China, Colombia, Costa Rica, Cuba, Czechoslovakia, Denmark, Dominican Republic, Ecuador, Egypt, El Salvador, Ethiopia, France, Greece, Guatemala, Haiti, Honduras, India, Iran, Iraq, Lebanon, Liberia, Luxembourg, Mexico, Netherlands, New Zealand, Nicaragua, Norway, Panama, Paraguay, Peru, Philippines, Poland, Saudi Arabia, South Africa, Syria, Turkey, Ukraine, Union of Soviet Socialist Republics, United Kingdom of Great Britain and Northern Ireland, United States of America, Uruguay, Venezuela, and Yugoslavia.

Perhaps you'd like to contemplate what these countries have in common. Or should I say *had* in common, as they're not all around today, or in some cases still around but under different names. (See Answers.)

▼

AT the right is a path that starts at the origin and ends up six units to the right, with the only legal steps being one unit to the right, one unit up diagonally, or one unit down diagonally. There are a total of 51 such paths. In math jargon, 51 is the sixth Motzkin number, named for Berlin-born American mathematician Theodore Samuel Motzkin (1908–1970).

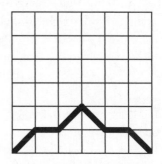

Motzkin numbers arise in a variety of contexts within the field of combinatorics, the snazzy word for "counting." For example, Motzkin numbers count how many expressions can be created by combining a letter with nested parentheses. In the case of 6 total characters, this gives us a more compact way of listing the 51 possibilities:

You can convert to paths using the scheme (= /, x = — , and) = \.

xxxxxx	xxxx()	xxx()x	xxx(x)	xx()xx	xx()()
xx(x)x	xx(xx)	xx(())	x()xxx	x()x()	x()()x
x()(x)	x(x)xx	x(x)()	x(xx)x	x(xxx)	x(x())
x(())x	x(()x)	x((x))	()xxxx	()xx()	()x()x

```
()x(x)    ()()xx    ()()()    ()(x)x    ()(xx)    ()(())
(x)xxx    (x)x()    (x)()x    (x)(x)    (xx)xx    (xx)()
(xxx)x    (xxxx)    (xx())    (x())x    (x()x)    (x(x))
(())xx    (())()    (()x)x    (()xx)    (()())    ((x))x
((x)x)    ((xx))    ((()))
```

52 $\left[\, 2^2 \times 13 \,\right]$

BESIDES being famous as the number of weeks in a year and the number of white keys on a piano, 52 shows up in all sorts of games. The most obvious connection is that there are 52 cards in a deck, consisting of 4 suits with 13 cards apiece. Today those suits are conventionally spades, hearts, diamonds, and clubs, but many other symbols have appeared over the centuries, among the earliest known being the polo sticks, coins, swords, and cups of the fourteenth century.

▼

LESS familiar, perhaps, is the legend surrounding Parker Brothers' initial rejection of Monopoly. As the story goes, game developer Charles Darrow was informed that his creation contained "52 fundamental errors," among them the complexity of the rules and the absurdly long playing time. The happy ending didn't occur until Darrow succeeded in selling the game at Wanamaker's Department Store in Philadelphia, whereupon Parker Brothers reconsidered and made Darrow the first person to become a millionaire via the game-invention route.

▼

THERE were 52 squares in the Aztec game Patolli. As it happens, the Aztec calendar, like the Mayan calendar, operated on a 52-year cycle, but that may be just coincidental, as Patolli predates the Aztec civilization. Patolli is said to have had religious significance. While avoiding the human sacrifice ascribed to the infamous Aztec ball game *ullamaliztli*, Patolli players could in theory wind up as indentured servants if they bet more than they can afford. (No doubt this would qualify as a "fundamental error" in the view of Parker Brothers.)

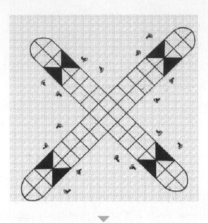

MOST people are familiar with the structure of a limerick: five lines, with the first one rhyming with the second, the third and fourth forming a different rhyming pair, and the fifth line rhyming with the first. This AABBA pattern is but one of 52 rhyme schemes that can be created from a poem of five lines. In the world of combinatorics, 52 is known as a Bell number, after Scottish-born mathematician and author Eric Temple Bell.

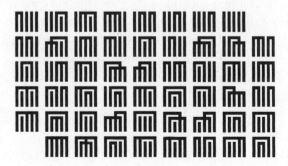

The Bell numbers, like the Motzkin numbers introduced in **51**, can be described in a number of different but equivalent ways. The figure above lists the 52 possible configurations of Genjiko, a Japanese art form whose icons consist of five columns joined by a horizontal bar at the top. If you think of a column as a line of poetry and a bar joining two columns as meaning that those lines rhyme, the 52 possible rhyme schemes for a five-line poem are thereby delineated.

The list of Bell numbers can be created by a process that evokes both Pascal's Triangle and the Fibonacci numbers. Basically you start with the assumption that 1 is a Bell number, at which point you make a little triangle like so, adding the numbers of the top row to form the number below it. Since $1 + 1 = 2$, you welcome 2 to the list.

Now that 2 has joined the club of Bell numbers, it gets added to the top row and is now used to generate the next Bell number, 5, as follows:

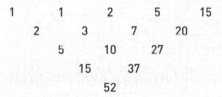

Two more steps of this same process produce the triangle above, which identifies 52 as the fifth Bell number.

In math-speak, the nth Bell number is the number of partitions of a set of n objects or the number of ways in which n distinguishable balls can be placed into n indistinguishable urns. The nth Bell number is also the number of multiplicative partitions of a number with n distinct prime factors. In particular, the number $2310 = 2 \times 3 \times 5 \times 7 \times 11$ can be written as a product in precisely 52 ways.

▼

IN 1917, British puzzlemaster Henry Dudeney introduced the No-Three-in-a-Line Problem, which asked how many dots could be placed on an $n \times n$ lattice without ever having three dots lying on the same straight line. It is not hard to see that the theoretical maximum for the $n \times n$ case is $2n$ dots, because anything beyond that would produce three dots in some row or column. For what values of n is the maximum of $2n$ actually achieved? Well, it has long been conjectured that for sufficiently large n, no solution exists. At this writing, the largest known solution, depicted below, is the 52-unit construction discovered in 1992 by the colorfully named German mathematician Achim Flammenkamp.

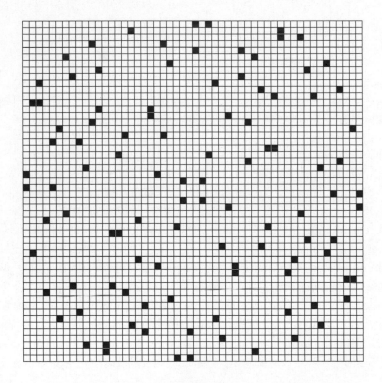

53 [prime]

53 = (3 × 16) + 5, so 53 = 35 in the hexadecimal system (base 16). No other two-digit number reverses itself upon hexadecimal conversion. But when you allow for more digits, reversals are spawned by 371 (= 173_{16}), 5141 (= 1415_{16}), and 99481 (= 18499_{16}), and all three of these numbers are multiples of 53.

▼

VENICE 53 was the name given to the gambling affliction that hit Italy in 2004, when the number 53 failed to appear in any of Venice's biweekly lotto drawings for an inordinate period of time (eventually over 180 drawings between May 2003 and February 2005). Gamblers lost over $2 billion betting on the reemergence of 53, apparently forgetting that despite 53's prolonged drought, its chances of coming up on any particular drawing were no greater than those of any other number.

▼

48 49 50 51 52 53 54 55 56 57 58

53 is the smallest prime whose five neighbors in either direction are all composite. (The next primes with this property are 89 and 157.)

The frequency of primes within the positive integers is a subject that has been explored for centuries. In the early 1800s, Carl Friedrich Gauss and French mathematician Adrien-Marie LeGendre (who had proved Fermat's Last Theorem for the single case $n = 5$), both conjectured that the number of primes less than n, denoted $\pi(n)$, was approximately $\frac{n}{\ln(n)}$, where $\ln(n)$ is the natural logarithm of n. This result, now known as the Prime Number Theorem, was proved independently by Hadamard and Vallee Poussin in 1896.

▼

LONGTIME baseball fans will recall that Don Drysdale wore number 53 for the Los Angeles Dodgers in the 1960s, but not many are aware that his number was intentionally passed on to Herbie, the Volkswagen Beetle that first appeared in the movie *The Love Bug*, by producer and Dodgers fan Bill Walsh.

54 $\left[\, 2 \times 3^3 \,\right]$

THE area of a regular pentagon equals $(\frac{5}{4})a^2\tan(54)$, where a is the length of a side. The 54 arises because the area is calculated by dividing the pentagon into five congruent isosceles triangles, each with angle measures of 54°, 54°, and 72°.

THE most famous disco of them all, Manhattan's Studio 54, was located at 254 West 54th Street.

THERE are a total of $6 \times 9 = 54$ faces on a Rubik's cube.

THE expression "54–40 or fight" was a campaign slogan of James K. Polk in 1844. The slogan's implicit threat was that Polk would go to war with Canada in order to set the northern border of Oregon at the parallel 54°40′. It didn't happen, and the border was officially set at 49° in 1846. The 54°40′ parallel lives on, however: It is the southernmost latitude of Alaska, and it is the patriotic and surprising name of a popular quilt pattern.

OFFICERS Gunther Toody and Francis Muldoon drove a red car in *Car 54*,

Where Are You? whereas actual police cars of the era were painted green. The car's distinctive coloring enabled it to be driven on location in New York City without creating confusion. The show's viewers never knew the difference, because *Car 54* (1961–63) was broadcast only in black and white.

▼

AT this writing, Africa has 54 countries, from Algeria to Zimbabwe.

▼

IN our discussion of quadratic forms in **15**, we mentioned that the great Indian mathematician Srinivasa Ramanujan identified 54 expressions of the form $aw^2 + bx^2 + cy^2 + dz^2$ that can generate all positive integers. The list is as follows:

(1,1,1,1)	(1,1,1,2)	(1,1,1,3)	(1,1,1,4)	(1,1,1,5)	(1,1,1,6)
(1,1,1,7)	(1,1,2,2)	(1,1,2,3)	(1,1,2,4)	(1,1,2,5)	(1,1,2,6)
(1,1,2,7)	(1,1,2,8)	(1,1,2,9)	(1,1,2,10)	(1,1,2,11)	(1,1,2,12)
(1,1,2,13)	(1,1,2,14)	(1,1,3,3)	(1,1,3,4)	(1,1,3,5)	(1,1,3,6)
(1,2,2,2)	(1,2,2,3)	(1,2,2,4)	(1,2,2,5)	(1,2,2,6)	(1,2,2,7)
(1,2,3,3)	(1,2,3,4)	(1,2,3,5)	(1,2,3,6)	(1,2,3,7)	(1,2,3,8)
(1,2,3,9)	(1,2,3,10)	(1,2,4,4)	(1,2,4,5)	(1,2,4,6)	(1,2,4,7)
(1,2,4,8)	(1,2,4,9)	(1,2,4,10)	(1,2,4,11)	(1,2,4,12)	(1,2,4,13)
(1,2,4,14)	(1,2,5,6)	(1,2,5,7)	(1,2,5,8)	(1,2,5,9)	(1,2,5,10)

A word on the notation seems appropriate. When we say that (1,2,3,8) is a universal quadratic form, we mean that any positive integer can be expressed in the form $w^2 + 2x^2 + 3y^2 + 8z^2$ for some choice of $w, x, y,$ and z, and similarly for the others. My personal favorites are (1,2,3,4) (the first four positive integers), (1,2,3,6) (the first three followed by their product or sum), (1,2,3,5) (Fibonacci numbers), and (1,1,2,6) (my birthday, if the first and third commas are removed and US birthday notation is used).

Ramanujan's list was remarkable given that he had no formal training in mathematics and no access to modern calculating tools. Ramanujan died in 1920 at the age of 32.

55 [5 × 11]

55 is the tenth Fibonacci number as well as the tenth triangular number and the largest number that is both Fibonacci and triangular.

▼

```
        A B B A C A C B B A        A C A B C A C C A B
         C B C B B B A B C          B B C A B B C B C
          A A A B B C C A            B A B C B A A A
           A A C B A C B              C C A A C A A
            A B A C B A                C B A B B A
             C C B A C                  A C C B C
              C A C B                    B C A A
               B B A                      A B A
                B C                        C C
                 A                          C
```

THE two triangles above each contain 55 letters and are created by a method that yields some surprising results. To begin with, the letters A, B, and C are placed in the top row more or less arbitrarily. The second row is determined by the following rule: If two consecutive letters in the first row are different, the third letter is placed between them in the second row. If two consecutive letters are the same, that letter is repeated below. The process continues until you are left with a single letter in the tenth row.

Note that in the left-hand triangle, the first and last entries of the top row are the same, namely A, and the letter at the bottom is also an A. In the right-hand triangle, the top left and top right letters are different (A and B), and the letter at the bottom is the third one (C). Guess what? No matter how you place the letters in the top row, you will always get the same result! (This property turns out to be a function of the triangle's height: It works for any triangle whose height is an even number not divisible by 4.)

▼

THE number 55 also has an odd but logical significance within the game of basketball. Historically, 55 was the highest uniform number a player could have without special dispensation. Why? Check out the next time you see a referee call a personal foul. The ref must inform the scoring table who committed the foul, and the standard method to do that is to signal that player's uniform number by holding up the appropriate number of fingers on each hand. Until refs start growing extra fingers, the digits 6, 7, 8, and 9 are out of the question, so the biggest number they can comfortably signal is 55.

NOT only is 55 a triangular number, it is also square-pyramidal. What does that mean? Well, start with a 5 × 5 array of bowling balls. Using the spaces between the balls, place a 4 × 4 array on top of that, and so on. By the time you reach the single bowling ball at the top of the pyramid, you will have used a total of 25 + 16 + 9 + 4 + 1 = 55 bowling balls. (Otherwise stated, a 5 × 5 square grid forms a total of 55 squares along its lattice lines.)

WHAT'S with this peculiar shape?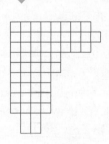

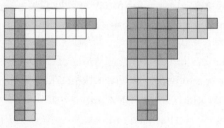

The answer is that the shape consists of 55 individual blocks and can be tiled either by using rectangles of sizes 1 × 1 through 1 × 10 or by using squares having sizes 1 × 1 through 5 × 5. This simultaneous tiling is a two-dimensional way of illustrating that 55 is both triangular and square pyramidal—a rare animal indeed, as we will discover in **91**.

56 $\left[\ 2^3 \times 7\ \right]$

IN the United States, the single biggest reason why the number 56 is part of the public consciousness has to be Joe DiMaggio's 56-game hitting streak. DiMaggio's achievement came in 1941 and is now widely viewed as unassailable, right up there with Cy Young's 511 wins, Sam Crawford's 312 triples, and, yes, Cy Young's 316 losses. No one has ever seriously challenged DiMaggio's mark, the best hitting streak since 1941 being Pete Rose's 44-game streak in 1978. John Allen Paulos has pointed out that the *a priori* likelihood of DiMaggio, a career .325 hitter, hitting safely in 56 straight games over the course of a season is exceedingly small—on the order of one in a hundred thousand.

▼

THE 56 Aubrey holes of Stonehenge would appear to be much longer-lived than DiMaggio's streak, but in some sense that's not the case. Stonehenge itself dates back to 3000 BC, but the holes weren't formally registered until antiquarian John Aubrey discovered them on a visit in 1666, and most weren't excavated until the twentieth century, if at all. The claim that the Aubrey holes have an astronomical function isn't backed by scientific consensus.

▼

THE names of how many future US presidents can be found among the 56 signers of the Declaration of Independence?

If you ever get hit with that trivia question, phrased just so, a quick scan of the list of signers indicates that the answer is four. Two of them (John Adams and Thomas Jefferson) actually became president. Another (Benjamin Harrison) was the father of William Henry Harrison, the ninth president, and shared a name with that Harrison's grandson, who became the twenty-third president. The fourth and final name is a bit of a stretch, but New Hampshire's Josiah Bartlett qualifies in some sense, as his name (minus one of the trailing *t*'s) was used in the television series *The West Wing* as that of the fictional president played by Martin Sheen.

1	2	3	4	5
2	3	4	5	1
3	4	5	1	2
4	5	1	2	3
5	1	2	3	4

THE above is the simplest possible example of a 5 × 5 Latin square. As we saw in **36**, a Latin square of order n is an $n \times n$ array, typically using the numbers 1 through n, such that no row or column repeats a number. Once you have created such a square, you can always rearrange the columns so that the first row consists of the numbers 1 through n in that order. And once you've done that, you can always rearrange the rows so that the first column is 1 through n reading down. The result is a called a reduced (or normalized) Latin square. The above square is kind of a self-working reduced square, but altogether there are 56 reduced squares, listed below in four groups of 14. The total number of 5 × 5 Latin squares equals 56 × 5! × 4! = 161,280. In general, if R_n equals the number of reduced Latin squares of order n, the total number of Latin squares of order n equals $R_n \times n! \times (n-1)!$

12345 12345 12345 12345 12345 12345 12345 12345 12345 12345 12345 12345 12345 12345
23451 21453 21534 21534 21453 21453 21534 21534 23451 23451 23514 23514 23154 23154
34512 34512 34152 34251 35214 35214 35421 35412 31524 31524 31452 31452 34512 34521
45123 45231 45213 45123 43521 43521 43152 43251 45132 45213 45123 45231 45231 45213
51234 53124 53421 53412 54231 54132 54213 54123 54213 54132 54231 54123 51423 51432

12345 12345 12345 12345 12345 12345 12345 12345 12345 12345 12345 12345 12345 12345
21453 23514 23514 23514 23154 23154 23451 23451 23451 23514 24153 24513 24531 24531

34521 34152 34251 34251 35412 35421 35124 35214 35214 35421 31524 31254 31254 31452
45132 45231 45123 45132 41523 41532 41532 41523 41532 41253 45231 45132 45123 45123
53214 51423 51432 51423 54231 54213 54213 54132 54123 54132 53412 53421 53412 53214

12345 12345 12345 12345 12345 12345 12345 12345 12345 12345 12345 12345 12345 12345
24513 24531 24153 24153 24531 24531 24153 24153 24153 24513 24531 24513 25134 25413
31452 31452 35214 35421 35124 35412 35214 35412 35421 35124 35214 35421 31452 31254
45231 45213 41532 41532 41253 41253 43521 43521 43512 43251 43152 43152 43521 43521
53124 53124 53421 53214 53412 53124 51432 51234 51234 51432 51423 51234 54213 54132

12345 12345 12345 12345 12345 12345 12345 12345 12345 12345 12345 12345 12345 12345
25431 25413 25431 25413 25134 25134 25134 25431 25413 25431 25134 25134 25413 25413
31254 31524 31524 31524 34251 34512 34521 34152 34251 34512 34251 34512 34152 34512
43512 43152 43152 43251 41523 41253 41253 41523 41532 41253 43512 43251 43521 43152
54123 54231 54213 54132 53412 53421 53412 53214 53124 53124 51423 51423 51234 51234

57 $\left[\ 3 \times 19 \qquad 2^5 + 5^2\ \right]$

SHOWN at right is what is described as the most complex
Chinese character still in use. It apparently involves 57
separate pen strokes (though I confess I haven't counted)
and represents *biang*, as in biang-biang noodles.

▼

IN the United States, the number 57 is better known in food circles through
H. J. Heinz's "57 varieties." The company actually had well more than 57 offerings when the slogan was introduced in 1896, but 57 lives on in their
corporate culture. The address of Heinz's world headquarters in Pittsburgh,
Pennsylvania is P.O. Box 57.

About the only way in which Heinz didn't exploit the number 57 was to
build an office building with 57 floors. That task was accomplished by F. W.

Woolworth in 1913, and the Woolworth Building was the world's tallest until surpassed in 1930 by both 40 Wall Street and the Chrysler Building.

IF you had three different colors at your disposal and you had to paint each face of a cube with one of those three colors, you could create 57 distinct cubes, where *distinct* in this context means that no coloring could lead to another via rotation. In particular, the number of possible colorings using k colors (with $k = 1$ through 6) can be shown to equal $\frac{(k^6 + 3k^4 + 12k^3 + 8k^2)}{24}$, where 24 just might be recognizable as the number of possible rotations of the cube (six faces each rotated in one of four ways). Plugging in $k = 3$ yields the desired 57 colorings.

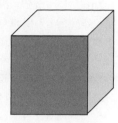

IN advanced trigonometry and calculus, angles are generally measured in radians rather than degrees. A radian is defined as the angle measure that makes an arc of a circle equal to its radius, as in the diagram: It turns out that one radian equals slightly more than 57 degrees. The exact number is $\frac{360}{(2\pi)}$, which to three decimal places equals 57.296.

58 [2 × 29]

THE game Hexxagon, an online and/or arcade variation of Othello, is played on a hexagonal grid with three of the interior flattened hexagons blacked out, for a total of 58 usable spaces. (Game spaces are usually called squares, but that seemed inappropriate here.)

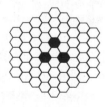

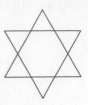

REGULAR polygons such as the hexagon can be extended to form different shapes, known as stellations of the original object, and a little time with stellations will produce another appearance of the number 58. We start with the observation that the hexagram to the right, usually thought of as an overlapping pair of equilateral triangles, is also just a regular hexagon with its edges extended until they meet:

A regular octagon (below left) has not one but two stellations, as the edges in the original extension (below middle) can themselves be extended to form the figure on the far right.

IN three dimensions, it is impossible to extend a tetrahedron or a cube, just as it is impossible to extend a triangle or square in two dimensions. However, each of the other five Platonic solids (see **5**) has at least one stellation. An octahedron (eight faces) has one stellation, known as the stella octangula. The dodecahedron (12 faces) has three distinct stellations. Finally, the icosahedron (20 faces) has a preposterous total of 58 stellations.

59 [prime]

IF you multiply the first two primes and add 1, you get 7, a prime. If you multiply the first three primes and add 1, you get 31, a prime. We can keep going in this manner:

$$2 \times 3 + 1 = 7 \qquad \text{prime}$$
$$2 \times 3 \times 5 + 1 = 31 \qquad \text{prime}$$
$$2 \times 3 \times 5 \times 7 + 1 = 211 \qquad \text{prime}$$
$$2 \times 3 \times 5 \times 7 \times 11 + 1 = 2311 \qquad \text{prime}$$

But if we went one more step, the pattern would come to a screeching halt, and the number 59 makes an appearance:

$$2 \times 3 \times 5 \times 7 \times 11 \times 13 + 1 = 30{,}031 = 59 \times 509$$

Of course, there's no earthly reason why the product of the first n primes plus 1 should itself be prime. However, this type of construction is historically important because a variant of it was used by Euclid, circa 300 BC, to show that there must be an infinite number of primes. To wit, suppose that there were only finitely many primes. Multiply those primes together and add 1. This new number may or may not be prime, but we know for certain that none of its prime factors could be in our supposedly complete list, because no number other than 1 divides evenly into two consecutive numbers. Therefore any finite set of primes cannot possibly be sufficient.

▼

THE number 59 also produces an interesting chart when you start dividing it by small numbers, starting as small as possible.

When you divide 59 by	You get a remainder of
2	1
3	2
4	3
5	4
6	5

You will note that the above table works because 59 is one less than 60, the smallest number that is evenly divisible by 2, 3, 4, 5, and 6. (Divisibility by 6 of course follows from divisibility by 2 and 3.) With that in mind, you should be able to come up with a number that when substituted for 59 enables the table to be extended to include division by 7. (See Answers.)

60 [$2^2 \times 3 \times 5$]

THERE wasn't much to say about the numbers 58 or 59, but with 60 comes a wealth of connections. The key is divisibility. Whereas 58 had but two prime factors and 59 was prime, 60 is divisible by each of the first three primes and is in fact the smallest number divisible by each of the first six positive integers and the smallest number with 12 factors: 1, 2, 3, 4, 5, 6, 10, 12, 15, 20, 30, and 60 itself.

▼

EACH of the three angles of an equilateral triangle measures precisely 60 degrees.

▼

THERE are 60 seconds in a minute and 60 minutes in an hour. These properties are taken for granted today but they date back to Mesopotamian civilizations, which used 60 as the base for their number systems. The Babylonians had no symbol for zero, but they were able to generate the first 59 integers using only two basic symbols.

▼

HERE'S a slightly more technical property of the number 60. It starts with the observation that there are 120 permutations of five elements (120 = 5! = 5 × 4 × 3 × 2). Any permutation is built up of "transpositions," meaning the switching of two particular elements. For a given permutation, either an even or an odd number of transpositions will be required. (That sounds silly, but the point is that there is more than one way to build a permutation out of transpositions, but it is not possible for one way to involve an even number and the other an odd number.) Of the 120 total permutations of five elements, 60 are "even." The permutation shown to the right is even because it involves an even number (namely two) of individual transpositions—B and E switch with one another, as do C and D.

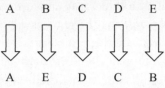

Because a combination of even permutations is itself even, in mathematical terms the even permutations form a subgroup of the full "symmetric group" of permutations of five elements, denoted S_5. This subgroup, called A_5, is the smallest possible non-abelian simple group. The non-abelian part means that multiplying different permutations isn't commutative, while the simple part means that the group contains no proper normal subgroups other than the identity permutation.

If the foregoing didn't make sense, don't worry. It's just math talk, cleverly designed to scare people away. But as far as the math world is concerned, it's an important appearance of the number 60.

▼

TO close on a more understandable note, the diagram shown is the board for Bridg-It, a game devised by the late American mathematician and economist David Gale circa 1960. The board's 60 dots are raised and serve as landing places for pieces of two different colors (red and black in the actual game), the objective being to create an uninterrupted path

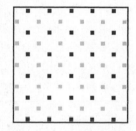

of your chosen color from one side of the board to the other. It is not possible for the game to end in a draw: one player or the other will always be able to create such a path. (By way of contrast, in the presidential election of 1960, John F. Kennedy won despite being unable to fashion either a north-south or east-west chain of states. He could have gone north-south had he been able to jump over Lake Michigan—but if that had been possible, Richard Nixon would have had an East-West chain to go with his multiple north-south "victories." In any event, no "jumping" is permissible in Bridg-It.)

▼

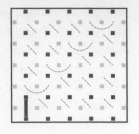

SOMEWHERE in my house there's an old game of Bridg-It that I played as a kid. (I was about 6 years old when the game was invented.) But I was never aware that the game is trivial from a mathematical standpoint, in that the first player can always force a win. The strategy, as deduced by Oliver Gross and communicated by Martin Gardner back in the '60s, is for black to place his first move as shown below, and then to follow any move by his opponent with a move that touches the other end of the dotted line touched by the opponent's piece.

While the triviality of Bridg-It was no doubt a disappointment to its inventor and manufacturers (the small Rhode Island firm of Hassenfeld Brothers, now the not-so-small company called Hasbro), it is not unusual for certain classes of games to be "determined" in one way or another. The mathematician who did the seminal work in this field was Ernst Zermelo (1871–1953), also a giant in the world of set theory. Zermelo is credited with the 1913 theorem that in chess, either white can force a win or black can force a win or either player can force a draw. As game theory developed, so did its lingo, and now you can say that "finite two-person zero-sum games of perfect information" are strictly determined. Fortunately for chess players, the so-called tree for chess is so vast that the actual calculation will always be beyond the capacity of mankind. But the Bridg-It example happens to be a case where the winning strategy can be mapped out explicitly.

61 [prime]

IN the TV game show *Jeopardy!*, the Jeopardy! and Double Jeopardy! rounds each consist of six categories with five questions (or answers) apiece. Throw in Final Jeopardy! and that's a total of $6 \times 5 + 6 \times 5 + 1 = 61$ questions and answers, although in most games the contestants don't have quite enough time to use up all of them.

THAT'S not the only connection between *Jeopardy!* and the number 61: In 1961 (the only year in the twentieth century that reads the same right-side up and upside down), a young newscaster named Alex Trebek made his TV debut for CBC, the Canadian Broadcasting Company. Trebek moved to the United States in 1973 to become a game-show host, and his talents were rewarded in 1984, when he was named emcee for the revived version of *Jeopardy!*

▼

SPEAKING of Canada, a hockey rink is 61 meters long, but with a catch. The 61 meters figure only applies to North American (i.e., National Hockey League) standards, and is an approximation, albeit a pretty good one, to the precise measurement of 200 feet. In the rest of the world, hockey rinks follow the metric system. Unfortunately for the number 61, the official length of international rinks is just 60 meters. But 61 has enduring significance in the world of hockey, as it represents the number of NHL scoring records owned or shared by the great Wayne Gretzky upon his retirement in 1999.

▼

WORKING south from Canada, Highway 61 makes its way from the Canadian border all the way to New Orleans. Early in its path it passes through Hibbing, Minnesota, the birthplace of Bob Dylan, who immortalized the road in his 1965 album *Highway 61 Revisited*.

▼

THE number 61 has served as an upper bound of sorts in US presidential elections, as no candidate has ever received more than 61% of the popular vote. At this writing, the top percentage vote getters in US election history were Warren Harding in 1920, Franklin Roosevelt in 1936, Lyndon Johnson in 1964, and Richard Nixon in 1972, with 60.5%, 60.6%, 60.6%, and 60.3% of the popular vote, respectively. Given that America begins and ends with an A, it is appropriate that A is the most common letter among the individual state names in the United States, with a total of 61 A's in the 50 state names. You may be wondering whether any presidential candidate has ever captured precisely those states with an A in them. The answer is no. Not even close.

▼

A standard eye chart contains 61 letters distributed among 11 rows. Such a chart is called a Snellen chart, in honor of Dutch ophthalmologist Hermann Snellen, who invented the chart back in 1862.

62 [2 × 31]

THE number 62 shows up in a well-known brainteaser. It starts with the figure below, which contains 62 squares.

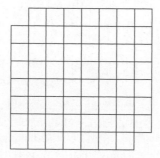

Now imagine that you have 31 dominoes, each measuring 1 × 2 squares, as follows:

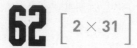

We know that 62 = 2 × 31, so the number of squares in the figure equals the total number of squares on the dominoes. The natural question, which you should feel free to tackle if you haven't seen it before, is whether it's possible to cover the entire 62-square region with the 31 dominoes. (See Answers.)

▼

SPEAKING of squares, 62 can be written as either $1^2 + 5^2 + 6^2$ or $2^2 + 3^2 + 7^2$. It is the smallest number with two different representations as the sum of distinct squares. And there are 62 different arrangements of five X's and four O's on a tic-tac-toe board that produce a win for X and X only.

63 $\left[3^2 \times 7 \right]$

DR. Subrahmanyam Karuturi has a long name. In fact, he has the longest name in the world. Not his own name, mind you, but the domain name below:

Iamtheproudownerofthelongestlongestlongestdomainnameinthisworld.com

If you take the time to count, you will discover 63 letters in the body of the URL. Dr. Karuturi's record cannot be broken (at least not without resorting to subdomain trickery), because 63 letters is the maximum allowable domain name with a .com extension.

▼

THE board for the Game of Goose consists of a picture of a goose covered by 63 discs. The game, a precursor to an extended family of dice/chase games, was originally registered by John Wolfe in 1597. The classic version by Laurie first appeared in 1831. That very same year saw the passage of the Game Act, which classified the pheasant, partridge, and grouse as game birds and defined a hunting season for them, but made no mention of geese. Have geese ever had a better year?

▼

25	16	80	104	90
115	98	4	1	97
42	111	85	2	75
66	72	27	102	48
67	18	119	106	5

91	77	71	6	70
52	04	117	69	13
30	118	21	123	23
26	39	92	44	114
116	17	14	73	95

47	61	45	76	86
107	43	38	33	94
89	68	63	58	37
32	93	88	83	19
40	50	81	65	79

31	53	112	109	10
12	82	34	87	100
103	3	105	8	96
113	57	9	62	74
56	120	55	49	35

121	108	7	20	59
29	28	122	125	11
51	15	41	124	84
78	54	99	24	60
36	110	46	22	101

PICTURED above are the five cross-sections of a 5 × 5 × 5 magic cube, in which the sums of every row, column, and diagonal are equal (to 315). Remarkably, the existence of a magic cube of order 5 wasn't proved until November 2003, when the pictured construction was revealed to the world by

Christian Boyer and Walter Trump. What *had* been known for some time that magic cubes of order 2, 3, and 4 were impossible, so a five-sided cube was the smallest possible nontrivial case. What also had been known was that if such a cube existed, its middle cube would have to be the number 63, the middle number in the string 1, 2, . . . 124, 125. Sure enough, that's the number in the center of Trump and Boyer's creation.

64 [2^6]

MANY of the special properties of the number 64 are the result of it being a power of two. In computing, where binary arithmetic is paramount, 64 bits is a well-established size for various data types. (The word *bit* is in fact shorthand for "binary digit.") And the 64 in *The $64,000 Question* was hardly a coincidence, as the number was attained by repeated doubling of a $1,000 wager, which itself was attained after successive doubling from $1 to $512 and then an approximate doubling to $1,000. (The proximity of 2 × 512 = 1,024 and 1,000 has caused confusion over the years in the computer world, where both numbers end up having a "K" attached to them, even though the former is a binary kilo and the latter a metric kilo.)

▼

THERE being six faces on a cube, the numbers on the doubling cube in backgammon range from $2^1 = 2$ to $2^6 = 64$.

▼

THE classic Chinese text *I Ching* (pronounced *e ching*) employs constructions called hexagrams, which are characters formed by stacking six bars, each row of which is either broken or unbroken (yin/yang). Those familiar with binary logic will at once recognize that there are $2^6 = 64$ possibilities altogether.

▼

A better-known construction, though not widely known for its ties to the number 64, is Braille. There are 64 characters in Braille, each formed by placing a raised dot (or not) in one of six fixed locations. As with the *I Ching*, the result is 64 different possibilities, though there are different versions of Braille. Only 26 of the characters are devoted to letters, with the first ten letters doubling as digits, subject to yet another character that warns you a number is coming.

The picture below is one of many ingenious mosaics made by computer graphics pioneer Ken Knowlton. This particular mosaic uses 16 copies of each of the 64 Braille characters to form a likeness of Helen Keller.

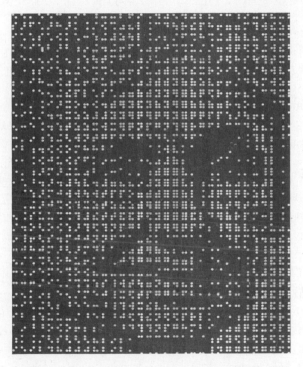

THE Beatles introduced "When I'm 64" in 1967, at which time no one in the group had even turned 30. For the record, Ringo Starr turned 64 in July 2004, while Sir Paul McCartney came of age in June 2005.

CHESS champion Bobby Fischer (1943–2008) made it to age 64 but no further, a poetic lifespan in that a chessboard has 64 squares. The alternating black/white pattern of the chessboard actually provides a clue for the classic puzzle introduced in **62**.

65 $\left[\, 5 \times 13 \,\right]$

IN some ways the number 65 is a relic of the twentieth century. For years it stood as the mandatory retirement age, but that was when people didn't live as long or work as long as they do today. And 65 miles per hour is perhaps the most familiar speed limit for seasoned drivers, but somehow it doesn't seem likely to survive the latest century and in fact has already fallen in a number of states. Fortunately, American society found a new way to honor the number 65 in 2001, when the NCAA basketball tournament was expanded from 64 to 65 teams. The idea was that a "play-in" game would be held during the week prior to the official tournament opening, thus reducing the field back to 64 again.

▼

MATHEMATICALLY speaking, the strange thing about the number 65 is that even though it is one away from being a square, several of its best-known properties revolve around squares. We'll get things rolling by the casual observation that 65 minus 56 (its reversal) equals 9, a square, while 65 plus 56 equals 121, also a square.

▼

BETTER known is the fact that 65 is the smallest number that can be represented as the sum of two squares in two different ways: $65 = 8^2 + 1^2 = 4^2 + 7^2$. Geometrically, this equation can be brought to life by the diagram on the next page, in which one line segment is the diagonal of a 1×8 rectangle and the other segment is the diagonal of a 7 × 4 rectangle. The two line segments appear to be about the same length, and any lingering doubt is removed by the Pythagorean Theorem, which shows that the length of each segment equals the square root of 65.

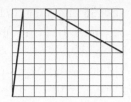

IN the same general area, sixty-five is the smallest number that can be a hypotenuse in four different ways, as shown by the identities $65^2 = 25^2 + 60^2 = 16^2 + 63^2 = 33^2 + 56^2 = 39^2 + 52^2$. The last of those four represents a right triangle with lengths 39, 52, and 65, which is just the famous 3-4-5 right triangle with each side multiplied by 13.

1	15	24	8	17
23	7	16	5	14
20	4	13	22	6
12	21	10	19	3
9	18	2	11	25

IN the magic square to the left, all rows, columns, and diagonals sum to 65. As we saw in our discussions of **15** and **34**, the magic constant for an $n \times n$ magic square is obtained by adding the numbers 1 through n^2 together and dividing by n: In this case, the sum of the numbers 1 through 25 equals 325, and dividing by 5 yields 65.

The magic square above is actually of a special variety. Note that the sum of the four corners of the square, coupled with the center square, equals $1 + 17 + 9 + 25 + 13 = 65$. The fun doesn't stop there. If you take any 3 × 3 subsquare and sum its corners plus center, you also get 65: The 3 × 3 square in the lower right yields $13 + 6 + 2 + 25 + 19 = 65$, and so on. Quite cool, I'm sure you'll agree, but in official math lingo the magic square is described using the more clinical adjectives of pandiagonal, associative, complete, and self-similar.

PERHAPS it is fitting to close the discussion of 65 and squares with a classic puzzle that claims 65 = 64. We start with the rectangle on the next page, consisting of 5 × 13 = 65 square units.

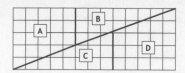

We now reassemble the four pieces to produce an 8 × 8 square—only 64 square units!

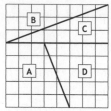

We know that some trickery is involved, being quite certain that 65 does not equal 64. The illusion results from the fact that the "diagonal" of the top rectangle is actually two line segments of different slopes. If we put the pieces of the square back into a rectangle, that rectangle would have a long, slivery hole in it of area 1.

▼

SPEAKING of holes, 8 × 8 boards, and the number 65, if you take an 8 × 8 chessboard and remove the middle 4 squares, you're left with 60 square units, exactly the number in a set of 12 pentominoes (see **12**). Back in 1958, Dana Scott of Princeton University showed that the 12 pentominoes could fill the chessboard-minus-hole space in precisely 65 different ways.

66 [2 × 3 × 11]

ROUTE 66 was once a major thoroughfare connecting Chicago with Los Angeles. The TV series of the same name was about the adventures of two men (Martin Milner and George Maharis) driving that route, which in essence represented the wide-open spaces of the western United States. The show was shot on location, but the locations seldom coincided with actual points on Route 66. (The Phillips 66 gas stations, named for the route, also showed up elsewhere.) As it happened, *Route 66* ran from 1960 to 1964, several years after the passage of the Interstate Highway Bill (1956) had already doomed the legendary road to oblivion.

▼

IN 6, we established that if you draw six dots on a piece of paper and connect every possible pair of dots with one of two colors, you are guaranteed to form a triangle whose sides are all the same color and whose vertices are at three of the dots. In **17**, we extended that concept to three colors. The extension to four colors gets a wee bit tricky, but it is possible to use the result from **17** to prove that if you connect all possible pairs from 66 dots with one of four colors, you are guaranteed to form at least one monochromatic triangle. The idea behind the proof is to connect 66 with 17, the magic number for three colors. So suppose you have a set of 66 dots such that every pair of dots connected with a line that is one of four colors: red, yellow, blue, or green. Pick a dot at random. There are 65 lines emanating from that dot, and since $65 > 16 + 16 + 16 + 16$, at least one of the four colors must appear 17 times. Without loss of generality we can assume that there are 17 red lines. Now consider the 17 dots at the other end of those red lines. If any two of them are joined by a red line, they must form a red triangle with the original dot as the third vertex. But if no pair is joined by a red line, then all the lines among those 17 dots must be blue, yellow, or green. But from the result of **17**, any set of 17 dots joined by line segments of three different colors must have a monochromatic triangle.

Note that the above proof doesn't demonstrate that 66 is the *minimal* number with the desired property. At this writing, that number is unknown, as is the case for the vast majority of higher-order Ramsey numbers, as they are called.

▼

THERE are 66 books in the Christian Bible. Tack on an extra 6, however, and everything changes: the number 666, also known as the number of the beast, is considered satanic.

A surveyor's chain, also called a Gunter's chain or just a chain, measures exactly $\frac{1}{80}$ of a mile, or $\frac{5,280}{80}$ = 66 feet. A patch of land one furlong in length and one chain in width measures precisely one acre. It is believed that the width of one chain corresponds to the width that could comfortably be plowed by a team of oxen. Though the measurement has declined in popularity, it can still be found today not so much as a width but as a length: A cricket pitch is 10 feet wide and 66 feet long.

67 [prime]

THE number 67 played a role in the Mansion of Happiness, America's very first board game. The game was manufactured by the W. & S.B. Ives Company, a stationer out of Salem, Massachusetts, beginning in 1843. Although the Puritan culture of the day frowned upon children wasting time with frivolous activities, Mansion of Happiness proved acceptable because of its moralistic overtones. Invented by clergyman's daughter Anne Abbott, the game encouraged good deeds as players worked their pieces through an inward spiral. Squares labeled charity, industry, and humanity represented rewards, while drunkenness and ingratitude were penalties. The objective of the game was to reach the Mansion of Happiness in the center of the board, square number 67.

A pizza can be divided into 67 pieces using only 11 straight cuts. In general, n cuts will divide a pizza into a maximum of $1 + \frac{n(n+1)}{2}$ pieces, or 1 plus the nth triangular number. The reason triangular numbers arise is that after the initial piece (the whole pizza), each successive cut adds a maximum of one piece, two pieces, three pieces, and so on.

▼

WITHIN the US Senate, any amendment to the Constitution must be approved by a two-thirds majority—67 of 100 senators present and voting. This same number applies to any change of Senate rules. In particular, in 1975, when the number of senators required to override a filibuster was reduced from 67 to 60, that change required the approval of 67 senators.

▼

"QUESTIONS 67 and 68" was a song on Chicago's first album (the group was then called the Chicago Transit Authority). The peculiar title generated many theories about its origin, but the eventual explanation of Robert Lamm (keyboard and vocals) was disappointingly mundane: The song was apparently about a man (possibly Lamm himself) whose girlfriend had asked a lot of questions during the prior two years. The above-referenced album came out in 1969, and "Questions 67 and 68," the fourth song on the first side, peaked at number 71 on the Billboard charts.

68 $\left[\, 2^2 \times 17 \, \right]$

HOW many unit discs can be placed inside a standard 8 × 8 chessboard? Even though there are 64 squares and the diameter of each disc equals the length of a square, the maximum is not 64. By nesting the discs together as shown in the diagram, it is possible to fit five columns of 8 and four columns of 7, for a total of $(5 \times 8) + (4 \times 7) = 68$ discs.

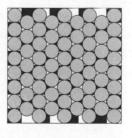

The packing of the discs in this manner is a hexagonal packing, much like the honeycomb found in **6**. The theoretical packing density of this pattern is approximately 90%, the most efficient packing using circles in the plane. The actual packing density in this case is $\frac{68\pi}{(4 \times 64)} \approx 83\%$, the lower figure arising because of the wasted space at the top and bottom of the board.

At this moment you're taking my word (and the suggestive drawing) that the nine columns actually fit inside the chessboard, but the fit isn't hard to prove. Note that the centers of the discs form the vertices of equilateral triangles spread across the grid. The total width of the discs equals the height of eight such triangles plus one (there is half a unit to the left and to the right of the triangles). An equilateral triangle whose sides are 1 has height equal to the square root of three (1.732) divided by two, so the total disc width is 1 + 4(1.732), or approximately 7.93—just barely under 8. In particular, no chessboard smaller than the standard 8 × 8 board would have accommodated the extra column.

▼

THE fraction $\frac{631,254}{314,526}$ equals 2 using octal (base 8) arithmetic. There are precisely 68 octal fractions in which the numerator and denominator consist of permutations of the digits 1 through 6 and whose quotient is an integer. Of perhaps more interest is that there are *no* solutions using the first five (or fewer) positive integers. The situation in base 10 is similar. If you take the first seven (or fewer) nonzero digits, there are no two permutations of those digits whose quotient is an integer. However, there are 2,338 solutions using eight digits, as in $\frac{86,314,572}{21,578,643} = 4$. Using all nine nonzero digits produces 24,603 distinct solutions.

▼

HOCKEY great Jaromir Jagr chose the uniform number 68 to honor the Prague Spring Rebellion of 1968, in which both of his grandfathers lost their lives. Jagr was the first Czechoslovakian player to be drafted by the NHL without first having to defect to the West.

69 [3 × 23]

STARTING with the earliest days of programmable calculators, the number 69 provided a somewhat mysterious upper bound. To illustrate, the Texas Instruments TI-59 calculator, introduced in 1977, could calculate any factorial from 2 to 69. Why that particular limit? Well, it turns out that 69 factorial—the product of all the positive integers less than or equal to 69—is approximately 1.71×10^{98}. Multiplying that number by 70 would push you beyond 10^{100}, the cutoff point for the TI-59 as well as the ensuing generation of pocket calculators—you have to stop somewhere, right? The result is that 69 gained a little fame as the largest number whose factorial could be calculated by a pocket calculator.

▼

THE number 69 also has the amusing property of being the only number whose square and cube contain all the numerals from zero to nine, once and only once.

$$69^2 = 4,761$$
$$69^3 = 328,509$$

▼

AN even more remarkable property of 69 arises from the standard alphanumeric coding, in which A = 1, B = 2, C = 3, and so on. Using this code, which is commonly found in the pseudoscience of numerology, the value of any word or collection of letters is defined as the sum of the individual letter values. In particular, this code can be applied to Roman numerals. Where 69 fits in is that its Roman numeral representation is LXIX. The value of these four letters in the standard code is equal to 12 + 24 + 9 + 24 = 69. Sixty-nine is one of only two numbers that are equal to their Roman numeral code values. Can you find the other (smaller) number? (See Answers.)

▼

IN the United States, Channel 69 is the highest-numbered UHF channel. Its supremacy dates back to 1982, when the Federal Communications Commission allotted channels 70 through 82 for cellular telephones.

▼

69 equals 105 in base 8, while 105 equals 69 in hexadecimal (base 16). Although this reversal is uncommon, it is shared by the numbers 64 through 69.

▼

THE Gordian's Knot puzzle shown to the right is named after a conundrum faced by Alexander the Great in 333 BC. Alexander ended up "untying" a knot, either by cutting it or removing it from its pin, after which conquering much of Asia was child's play. No cutting is allowed for the Gordian's Knot, but solving the puzzle involves a minimum of 69 moves.

70 $\left[\, 2 \times 5 \times 7 \,\right]$

THE proper divisors of 70 are 1, 2, 5, 7, 10, 14, and 35. The sum of these seven numbers exceeds 70 (it equals 74, actually), yet no subset of them adds up to precisely 70. Believe it or not, 70 is the smallest number with this property—that is, the smallest number that is less than the sum of its proper factors but which cannot be represented as the sum of a subset of those factors. It's even harder to believe that there's a name for this kind of number: a weird number. So 70 is the smallest weird number.

▼

TAKE a square, and divide it into eight isosceles right triangles, as follows:

By coloring each triangle black or white (and disregarding the lines as necessary), it is possible to create designs such as the following:

THE Izzi puzzle, designed by Frank Nichols and introduced by Virginia-based puzzle company Binary Arts (now called ThinkFun) in 1992, is made up of designs of precisely this type. How many are there altogether? Because each of the 8 triangles can be one of two colors, binary logic suggests that there should be a total of $2^8 = 256$ possible designs. But by that reckoning two designs such as

and

are counted separately, even though one can be converted into the other via a simple 90-degree rotation. Given that any one shape can be rotated four times before returning to its original state, it's tempting to just divide 256 by 4 to get 64, but that's not quite right, either, because of symmetries. The all-black design above, for example, never changes upon rotation.

Counting the actual number of designs is made easier by a mathematical tool called Burnside's Lemma. (It is named for the prolific nineteenth-century group theorist William Burnside, even though he apparently had nothing to do with this particular insight, which dates back to Cauchy and Frobenius in the first half of that century.) Although Burnside's Lemma is usually phrased in terms of "orbits" of "groups," here the idea is that you start with the total number of designs (256), add the number of designs that are symmetric with respect to 180-degree rotations (16), and then add the number that are symmetric with respect to 90-degree rotations in either direction (4 + 4), for a total of 280. *Now* you can divide by 4 (again, the number of possible rotations) to get a grand total of 70 distinct designs.

The idea behind Izzi is to arrange the pieces into various 8 × 8 squares such that the colors on any two adjacent pieces match up, as in the following:

But wait a minute. An 8 × 8 square uses 64 designs, and we just went to a lot of trouble to show that there are 70 designs altogether. What happened is that the folks at Binary Arts picked out six designs that wouldn't be used in the final puzzle. Those six designs are the ones shown at the beginning of this discussion.

71 [prime]

THE equation $7! + 1 = 71^2$ is one of only three known equations of its kind, the others being $4! + 1 = 5^2$ and $5! + 1 = 11^2$.

Just to be clear what's going on, let's spell those equations out:

$$4 \times 3 \times 2 + 1 = 25 = 5^2$$
$$5 \times 4 \times 3 \times 2 + 1 = 121 = 11^2$$
$$7 \times 6 \times 5 \times 4 \times 3 \times 2 + 1 = 5041 = 71^2$$

In other words, 71 is the largest known number whose square is one more than a factorial. That it happens to be one more than 7 factorial adds the finishing touch of having 7 and 1 on each side of the equation.

Research has indicated that $n! + 1$ is not a square for n up to one billion, and it is widely believed that 4, 5, and 7 are the only values of n that work. Technically, though, the problem (known in the trade as Brocard's Problem) remains very much in need of a proof.

▼

RECENT visitors to Hong Kong may have come across a little place called Club 71. The number 71 has special significance on the island because July 1, 1997, marked the handover of Hong Kong from the British to the Chinese. The date July 1 goes by the delightful name of *chat yat* in Cantonese.

Because 71 and 73 are both primes, 71 is a twin prime. And because 71 and 17 are both primes, 71 is an EMIRP, the name given to any prime that is the reverse of a prime. But the prime connections don't stop there. If you add the prime numbers less than 71, you get $2 + 3 + 5 + 7 + 11 + 13 + 17 + 19 + 23 + 29 + 31 + 37 + 41 + 43 + 47 + 53 + 59 + 61 + 67 = 568 = 8 \times 71$. For a number to be a factor of the sum of the primes less than it isn't exactly an important property, but it's a rare one: The next number to have it is 3,691,119. The final prime-related property is that $71^3 = 357,911$, a number obtained by stringing together five consecutive prime numbers beginning with 3.

▼

THE number 71 played a relatively brief role in a problem called the Happy End Problem. We start by noting that if you place four points on a piece of paper, it is not necessarily possible to connect them so as to create a convex quadrilateral, as follows:

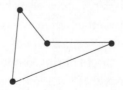

(*Convex* means that any line segment joining any two points inside the quadrilateral must lie entirely within the quadrilateral, so this particular figure doesn't qualify.)

However, if you add a fifth point, provided that no three of the points fall along the same line, it is always possible to connect four of the five points so as to create a convex quadrilateral.

The question obviously extends to polygons with more than four sides. Somewhere in this exploration, the legendary Paul Erdos demonstrated that 71 points will always be enough to guarantee the creation of a convex hexagon. His work actually proved that for any n, there is some number, called $g(n)$, such that $g(n)$ points would insure a convex polygon with n sides. He even gave an upper bound for $g(n)$ in a formula that yielded 71 when $n = 6$.

Precisely because Erdos's work was so general, other mathematicians had no trouble improving upon it, and today it is known that $17 \leq g(6) \leq 37$. But that's not the happy end. The reason this problem got its name is that two of Erdos's colleagues in working on the problem—Ester Klein and George Szekeres—became engaged during that process and eventually married.

72 $\left[\, 2^3 \times 3^2 \,\right]$

THE Rule of 72 is a shortcut that answers the question "How long will it take my money to double?" Suppose, for example, that you expect your investments to grow at an annual rate of 8%. Dividing 72 by 8 gives 9, and you can therefore expect your money to double after 9 years.

As it happens, I've chosen a case where the Rule of 72 estimate is almost exactly equal to the actual doubling time. The estimate will be less precise for very small or very large rates of return, but gives adequate perspective for most rates of return that could be considered reasonable. Other numbers can be used (you will see references to the Rules of 70, 71, and even 69.3, depending on the nature of the compounding), but 72 has the advantage of being evenly divisible by a wide range of numbers, among them

2, 3, 4, 6, 8, 9, 12, and 18. Remember, we're looking for an estimate, not an exact answer, so we can use whatever number suits our purposes.

The reason why the rule works at all for numbers in this range is that the equation for doubling times involves the natural logarithm of 2, which is very close to 0.69. Of course, if you want to quadruple your money instead of merely doubling it, just double the number that the Rule of 72 gives you. (A simple point, perhaps, but often overlooked.)

Perhaps the most remarkable aspect of the Rule of 72 is its longevity. It appeared in *Summa de Arithmetica* by Fra Luca Pacioli, an occasional collaborator of Leonardo da Vinci. The year was 1494. If Pacioli had invested his life savings into a perpetual investment paying $\frac{1}{7}$th of a percent each year, the Rule of 72 suggests that his investment would have doubled by the start of the twenty-first century.

SPEAKING of da Vinci and Pacioli, the two also collaborated in the production of the 72-sided sphere, a popular geometric rendering of the period. This image, spurred by Leonardo's drawings, appeared in Fra Pacioli's 1509 book *The Divine Proportion*. By approximating a sphere with polyhedra—whose volumes could be calculated—the "sphere" confirmed a result, known since Euclid's time, that the volume of an actual sphere was proportional to the cube of its radius. Today, the invention of integral calculus now long behind us, even high school students can derive the formula $V = (\frac{4}{3})\pi r^3$.

IN many international archery competitions, including the ranking rounds of the Olympic Games, archers shoot a total of 72 arrows. Because the inner gold of the bull's-eye is worth 10 points, a maximum score is 720. Although archery appeared in the first several Olympics, confusion over standardization knocked it out for many years. It returned for the Munich games of '72.

▼

WHEN *New York World* reporter Nellie Bly set off around the world on November 14, 1889, her goal was to match Phileas Fogg's fictional time of 80 days (see **80**). She did a bit better than that, as she returned to New York on January 21, 1890, 72 days and a few hours later.

▼

THE multiplicative properties of the number 72 plays a key role in a well-known puzzle that you are now invited to solve.

> A man in a bar is talking with the bartender. The bartender asks him if he has children and he replies that he has 3.
> He then asks their ages and the man responds that the product of their ages is 72.
> The bartender says, "That is not enough information."
> The man then says that the sum of their ages is on the front door of the bar.
> The bartender again says, "That is not enough information."
> The guy says, "My youngest likes ice cream."
> The bartender says, "In that case, I can figure out their ages."
> *What are the ages of the kids?* (See Answers.)

This puzzle is sometimes told using 36 instead of 72, but it is almost never told with the next number (after 72) that yields a solution. Care to guess what that number is? (See Answers.)

73 [prime]

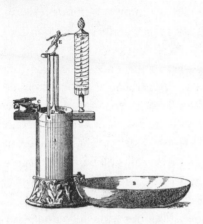

THE number 73 played a role in ancient clock making, a consequence of there being 365 days per year and the fact that $73 = \frac{365}{5}$. The clepsydra (water clock) pictured here is of the third century BC. Clocks with 5 water compartments and 73 teeth could create a 365-day cycle.

THE picture below demonstrates that 73 is a star number. There are a total of 73 dots in the star; the number of dots in the interior hexagon is 37, the reversal of 73.

THERE is an old puzzle that asks solvers to make various positive integers using four 4's, plus basic arithmetical operations, including decimal points, square roots, and factorials. For example, $1 = \frac{(4 + 4)}{(4 + 4)}$, $2 = \frac{4 \times 4}{(4 + 4)}$, and so on. Things get a bit more complicated for higher numbers, as you'd expect. For example, $70 = 44 + 4! + \sqrt{4}$, $71 = \frac{(4!+4.4)}{.4}$, and $72 = 44 + 4! + 4$. But I didn't include a simple expression for 73 because there isn't any. That's right. The number 73 is the smallest number for which no simple expression exists.

IN the context of Waring's Theorem, it can be said that all positive integers can be written as the sum of no more than 73 sixth powers (not necessarily distinct).

THE number 73 also shows up in a relatively recent theorem concerning different types of representation of integers. Recall from **4** that any positive integer can be expressed as the sum of at most four perfect squares. More generally, the famous 15 Theorem of Princeton's John Conway shows that if a quadratic form represents the numbers 1 through 15, it represents all positive integers. (A quadratic form is any expression in which the variables have degree two: $a^2 + b^2 + c^2 + d^2$ is such a form, but so is $a^2 + 2b^2 + 3c^2 + 4d^2$ and so on.)

Manjul Bhargava was introduced to these subtleties while a graduate student at Princeton, and he did them one better. One of his extraordinary results is that if a quadratic form (a positive-definite, matrix-defined form, but that's another matter) represents all prime numbers through 73, then it represents *all* prime numbers.

74 $\left[\, 2 \times 37 \,\right]$

FOR almost four centuries, the number 74 represented an unconfirmed limit for the three-dimensional sphere-packing problem. Johannes Kepler, in his famed 1611 treatise *Strena sue de nive sexangula* (on the six-cornered snowflake— see **6**), conjectured that no packing method is more efficient that the standard pyramidal (or hexagonal) stacking. The density of such a

stacking is $\frac{\pi}{3}\sqrt{2}$, just a tad over 74%, and the longstanding question was whether it was possible to find an alternative stacking method that occupied more than 74% of the available space.

In 1831, Carl Friedrich Gauss demonstrated that the Kepler conjecture was valid for *regular* lattices. Irregular lattices (it sort of means what you think it means) posed a problem for essentially two reasons. First, irregular packings are much more difficult to categorize than are regular packings. Second, it is in fact possible to construct asymmetrical arrangements that are denser than the hexagonal packing over a limited amount of space (the conjecture applied to the entirety of three-dimensional space).

Like the Four-Color Map Theorem, the Kepler conjecture went unresolved until the computer age provided the required analytical tools. University of Michigan's Thomas Hales issued a computer-driven verification in 1998. Kepler was apparently right all along.

It would have been fitting for Hales to have celebrated his achievement with a trip to Australia's Whitsunday Islands, located in the heart of the Great Barrier Reef. Though only seven of these islands are inhabited, they number 74 altogether.

▼

BACK in the Western Hemisphere, the Saffir-Simpson Hurricane Scale, developed by civil engineer Herbert Saffir and meteorologist Robert Simpson in the early 1970s, dictates that a storm doesn't officially get to be called a hurricane until its wind speed hits 74 miles per hour. This scale went into widespread use beginning in (you guessed it) 1974, when Simpson stepped down as director of the US National Hurricane Center.

75 $\left[\, 3 \times 5^2 \,\right]$

THERE are only 4! = 24 ways to rank four objects. However, if the ranking system allows for ties, that number increases all the way to 75. One way to list all 75 possibilities is to first list the standard 24 orderings of four objects, then group with brackets ties involving two players, then group ties involving three players, then strike through duplicates. The basic rule is that we remove any ordering in which the bracketed letters are not in alphabetical order. And, of course, we can't forget the seventy-fifth case—in which all four objects are tied. But something has gone wrong: the numbers in the columns don't add to 75 the way they should.

What possibilities have been left out? (See Answers.)

24

ABCD	ABDC	ACBD	ACDB	ADBC	ADCB
BACD	BADC	BCAD	BCDA	BDAC	BDCA
CABD	CADB	CBAD	CBDA	CDAB	CDBA
DABC	DACB	DBAC	DBCA	DCAB	DCBA

12

[AB]CD	[AB]DC	[AC]BD	[AC]DB	[AD]BC	[AD]CB
~~[BA]CD~~	~~[BA]DC~~	[BC]AD	[BC]DA	[BD]AC	[BD]CA
~~[CA]BD~~	~~[CA]DB~~	~~[CB]AD~~	~~[CB]DA~~	[CD]AB	[CD]BA
~~[DA]BC~~	~~[DA]CB~~	~~[DB]AC~~	~~[DB]CA~~	~~[DC]AB~~	~~[DC]BA~~

12

A[BC]D	A[BD]C	~~A[CB]D~~	A[CD]B	~~A[DB]C~~	~~A[DC]B~~
B[AC]D	B[AD]C	~~B[CA]D~~	B[CD]A	~~B[DA]C~~	~~B[DC]A~~
C[AB]D	C[AD]B	~~C[BA]D~~	C[BD]A	~~C[DA]B~~	~~C[DB]A~~
D[AB]C	D[AC]B	~~D[BA]C~~	D[BC]A	~~D[CA]B~~	~~D[CB]A~~

12

AB[CD]	~~AB[DC]~~	AC[BD]	~~AC[DB]~~	AD[BC]	~~AD[CB]~~
BA[CD]	~~BA[DC]~~	BC[AD]	~~BC[DA]~~	BD[AC]	~~BD[CA]~~
CA[BD]	~~CA[DB]~~	CB[AD]	~~CB[DA]~~	CD[AB]	~~CD[BA]~~
DA[BC]	~~DA[CB]~~	DB[AC]	~~DB[CA]~~	DC[AB]	~~DC[BA]~~

4

[ABC]D	[ABD]C	~~[ACB]D~~	[ACD]B	~~[ADB]C~~	~~[ADC]B~~
~~[BAC]D~~	~~[BAD]C~~	~~[BCA]D~~	[BCD]A	~~[BDA]C~~	~~[BDC]A~~
~~[CAB]D~~	~~[CAD]B~~	~~[CBA]D~~	~~[CBD]A~~	~~[CDA]B~~	~~[CDB]A~~
]DAB]C	~~[DAC]B~~	~~[DBA]C~~	~~[DBC]A~~	~~[DCA]B~~	~~[DCB]A~~

4

A[BCD] ~~A[BDC]~~ ~~A[CBD]~~ ~~A[CDB]~~ ~~A[DBC]~~ ~~A[DCB]~~
B[ACD] ~~B[ADC]~~ ~~B[CAD]~~ ~~B[CDA]~~ ~~B[DAC]~~ ~~B[DCA]~~
C[ABD] ~~C[ADB]~~ ~~C[BAD]~~ ~~C[BDA]~~ ~~C[DAB]~~ ~~C[DBA]~~
D[ABC] ~~D[ACB]~~ ~~D[BAC]~~ ~~D[BCA]~~ ~~D[CAB]~~ ~~D[CBA]~~

[ABCD]   1

TOTAL   75?

76 $\left[\, 2^2 \times 19 \,\right]$

IF you multiply 76 by itself, you get 5776, a number whose last two digits are 7 and 6. Obviously if you keep multiplying by 76, you always get a number that ends with 76. A number that is found at the end of all its powers is known as an automorphic number. Virtually all automorphic numbers end with either 25 or 76.

▼

WE have discussed (see **22**) partitions of integers into smaller positive integers. Within that subject, number theorists look into partitions of integers into primes—specifically, partitions into distinct primes. It turns out that small numbers have relatively few such partitions: For example, the number 15 has only two partitions into distinct primes—13 + 2 and 7 + 5 + 3. Eventually, of course, the number of distinct prime partitions of a number becomes much bigger than the number itself. It just so happens that 76 is the crossover point. There are 76 partitions of the number 76 into distinct primes, and here they are:

31+17+11+7+5+3+2	37+13+11+7+5+3	41+17+11+5+2	43+23+7+3	71+3+2 73+3
29+19+11+7+5+3+2	31+19+11+7+5+3	37+29+5+3+2	43+19+11+3	67+7+2 71+5
29+17+13+7+5+3+2	31+17+13+7+5+3	37+23+11+3+2	43+17+13+3	61+13+2 59+17
23+19+17+7+5+3+2	29+19+13+7+5+3	37+19+13+5+2	43+17+11+5	43+31+2 53+23
23+19+13+11+5+3+2	23+19+13+11+7+3	37+19+11+7+2	41+23+7+5	47+29
23+17+13+11+7+3+2	23+17+13+11+7+5	37+17+13+7+2		
	59+7+5+3+2	31+29+11+3+2		
	53+13+5+3+2	31+23+17+3+2		
	53+11+7+3+2	31+23+13+7+2		
	47+19+5+3+2	31+19+17+7+2	41+19+13+3	37+19+17+3
	47+17+7+3+2	31+19+13+11+2	41+19+11+5	37+19+13+7
	47+13+11+3+2	29+23+19+3+2	41+17+13+5	31+29+13+3
	43+23+5+3+2	29+23+17+5+2	41+17+11+7	31+29+11+5
	43+19+7+5+2	61+7+5+3		31+23+19+3
	43+17+11+3+2	53+13+7+3		31+23+17+5
	43+13+11+7+2	53+11+7+5	37+31+5+3	29+23+19+5
	41+23+7+3+2	47+19+7+3	37+29+7+3	29+23+17+7
	41+19+11+3+2	47+17+7+5	37+23+13+3	29+23+13+11
	41+17+13+3+2	47+13+11+5	37+23+11+5	29+29+17+11

77 $\left[\ 7 \times 11 \qquad 4 \times 4 + 5 \times 5 + 6 \times 6\ \right]$

THE Group of 77 is a coalition of developing nations that was first organized by 77 founding nations in 1964 at Algiers. The group has since expanded considerably, but its original purpose remains: to further the economic well-being of its member states.

During World War II, at the Sweden/Norway border, "77" was used as a password because its tricky pronunciation in Swedish made it easy to determine whether the speaker was Swedish, Norwegian, or German.

PARTITIONS OF 77

77 is the smallest number of coins that cannot be combined into a dollar. Note that a solution with 76 coins is trivial—75 pennies plus 1 quarter: 75 coins can produce a dollar via 70 pennies, 4 nickels, and 1 dime, and so on down to a silver dollar.

$$77 = 3 + 4 + 5 + 5 + 60 \qquad \frac{1}{3} + \frac{1}{4} + \frac{1}{5} + \frac{1}{5} + \frac{1}{60} = 1$$

NOTE the duplication of 5 in the above sums: It turns out that 77 cannot be written as the sum of *distinct* numbers whose reciprocals add to 1, and it is the largest number with that property. (See **22** and **96**.) For example:

$$100 = 2 + 6 + 7 + 8 + 21 + 56 \qquad \frac{1}{2} + \frac{1}{6} + \frac{1}{7} + \frac{1}{8} + \frac{1}{21} + \frac{1}{56} = 1$$

THE Rose Bowl in Pasadena, California, is the regular-season home of the UCLA Bruins and host to a bowl game of the same name since 1923. A *bowl*, by definition, has an unbroken sequence of rows instead of different tiers (so from any seat you can see every other seat), and the Rose Bowl has 77 numbered rows of seats from top to bottom.

78 $\left[\, 2 \times 3 \times 13 \,\right]$

A typical 15 × 15 crossword grid (the size used by daily newspapers in the United States) contains 78 entries. The number of entries can be quite a bit smaller, but puzzles with more than 78 entries are usually rejected by the top puzzle editors.

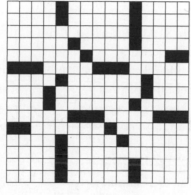

ELSEWHERE in the world of black squares, one of 35 possible heptominos (figures formed by joining seven squares at their edges) is shown below. Mathematicians take special interest in the tiling properties of such objects, and in 1989 Karl Dahlke proved the remarkable theorem that it takes at least 78 of the figures below to tile a rectangle.

SPEAKING of rectangles, a tennis court measures 78 feet from baseline to baseline.

THERE are 78 tarot cards in a complete deck—22 of the major arcana, and 56 of the minor arcana.

78 is the twelfth triangular number, being the sum of the numbers 1 through 12. From which we can see that there are a total of 78 presents (not including repetitions) in the song "The Twelve Days of Christmas."

THE figure to the right is known as Metatron's Cube, and is created by joining the centers of 13 circles arranged in a "Flower of Life" pattern. If you consider each such center-to-center connection to be a distinct line, it follows that the total number of lines in the diagram equals 13 (number of starting points) × 12 (number of ending points), divided by 2 (to avoid double-counting) = 78. Metatron was an angel in Judaism, and Metatron's Cube has a rich history of applications in alchemy and religion. At one time it was used as a means of warding off satanic powers.

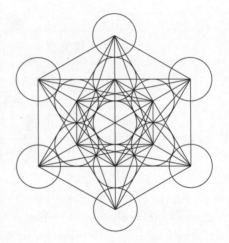

79 [prime $7 \times 9 + 7 + 9$ $2^7 - 7^2$]

NOT only are 79 and its reversal, 97, both primes, they form the sums of the following addition table, in which all of the numbers are reversible primes:

79 = 11 + 31 + 37
97 = 11 + 13 + 79

79 = 4 × 16 + 15 × 1 = 4 × 2⁴ + 15 × 1⁴

That's kind of a strange equation, but all it says is that 79 can be written as the sum of 19 fourth powers: 4 of two to the fourth and 15 of one to the fourth. In and of itself, that says nothing, because one aspect of Waring's Theorem assures us that *every* number can be written as the sum of 19 fourth powers. What makes 79 unusual is that it *requires* 19 fourth powers, and is the smallest number to do so.

▼

HERE'S a well-known brainteaser:

Three thieves stole a bunch of coconuts and agreed to divide them up evenly the following morning, then retired for the night. After an hour passed, the first thief decided to take his third. After dividing the nuts evenly, he had one left over, which he gave to a monkey. An hour later, the second thief did the exact same thing with the remaining nuts, as did the third thief an hour after that: In other words, both of the thieves took one-third of the coconuts they saw, and handed the single leftover coconut to a monkey. In the morning, the thieves divided the remaining nuts evenly and had one left over, which they gave to the monkey. What was the smallest number of coconuts they could have started with and not had to deal with coconut parts?

Can you show that the answer is 79? (See Answers.)

80 $\left[2^4 \times 5 \right]$

IN French, the number 80 is *quatre-vingts*, or literally four twenties. Perhaps the most celebrated appearance of this version of the number came in 1873, upon the publication of Jules Verne's *Le Tour de Monde en Quatre-vingts Jours*, otherwise known as *Around the World in Eighty Days*.

Technically, Phileas Fogg and his companion Passepartout didn't quite go "around the world." According to standards developed after Jules Verne's time, an official circumnavigation of the globe must pass through antipodal points—that is, two points that are completely opposite from one another. Here is Fogg's proposed itinerary:

London to Suez	rail and steamer	7 days
Suez to Bombay	steamer	13 days
Bombay to Calcutta	rail	3 days
Calcutta to Hong Kong	steamer	13 days
Hong Kong to Yokohama	steamer	6 days
Yokohama to San Francisco	steamer	22 days
San Francisco to New York	rail	7 days
New York to London	steamer	9 days
TOTAL		**80 days**

Of course, a series of unanticipated events derailed this schedule, but Fogg and Passepartout made it back to London almost a day early. The only problem was that they thought they were late: They had forgotten that by traveling west, they had crossed the international date line, and had therefore had picked up an extra day.

▼

ITALIAN economist Wilfredo Pareto (1848–1923) made the surprisingly powerful observation that 80% of his country's wealth was concentrated in 20% of the population. This phenomenon, dubbed Pareto's Law by legendary business thinker Joseph Juran, also goes under the name of the 80/20 Rule. Juran extended the rule to quality controls in manufacturing, where 80% of the problems could be attributed to 20% of the causes. The 80 and 20 aren't sacred, of course, but they form a useful guideline with plenty of applications. The table below contemporizes some of the old standbys. For a hypermodern version of Pareto's Law, one might use Woody Allen's famous observation: "Eighty percent of success is showing up."

THE 80/20 RULE, FILL-IN-THE-BLANKS STYLE

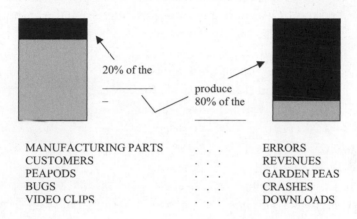

20% of the
_____
–

produce
80% of the
_____

MANUFACTURING PARTS	. . .	ERRORS
CUSTOMERS	. . .	REVENUES
PEAPODS	. . .	GARDEN PEAS
BUGS	. . .	CRASHES
VIDEO CLIPS	. . .	DOWNLOADS

81 [3^4]

ALTHOUGH Sudoku puzzles come in many sizes and shapes, a standard Sudoku has 81 squares. This size implicitly takes advantage of the fact that 81 is both a perfect square and a perfect fourth power.

▼

ALTHOUGH similar-looking puzzles appeared in the French daily *La France* in the nineteenth century (*carre magique diabolique*), it is generally acknowledged that the modern version of these puzzles was invented in 1979 by retired Indiana architect Howard Garns. Although Garns's efforts were published back then by Dell Magazines (as *Number Place*), the Sudoku craze didn't arrive in earnest until 2005.

▼

ANY odd perfect square > 1 can generate a Pythagorean triple by first representing that square as the sum of two consecutive integers—in this case, $81 = 40 + 41$. The triple (9, 40, 41)—where 9 is the square root of 81— can be seen to satisfy the usual $a^2 + b^2 = c^2$ equation. Of course, there's a much better-known famous separation of 81 into 40 and 41, and it goes like this:

> Lizzie Borden took an axe
> And gave her mother forty whacks.
> And when she saw what she had done
> She gave her father forty-one.

▼

THE eighth letter of the alphabet is H, and the first letter of the alphabet is of course A. Put the two together and you have the reason why 81 is used as an insignia by the Hell's Angels motorcycle club.

▼

FOR any four positive numbers p, q, r, and s, the following relation holds:

$$(p^2 + p + 1)(q^2 + q + 1)(r^2 + r + 1)(s^2 + s + 1) \, / \, pqrs \geq 81$$

By rewriting the left-hand side in a suitable way, can you prove the inequality? (Note that if p, q, r, and s all equal 1, you get $3^4 \geq 81$, which is actually an equality in this case.) (See Answers.)

82 [2 × 41]

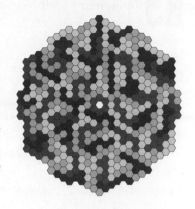

THERE are 82 different shapes (called hexahexes) that can be created by joining six hexagons along their edges. This diagram uses these 82 shapes to form a giant hexagon plus some trim. Note the nice touch of taking the unique hexahex with a hole in the middle and placing it in the center of the construction.

▼

A standard dartboard has a total of 82 regions (20 wedges with four regions apiece, plus the inner and outer bull in the center). The numbers around the circumference provide the scores for the corresponding wedges, and the credit (or blame) for their peculiar arrangement is generally given to nineteenth-century British carpenter Brian Gamlin. By placing small numbers around the desired big numbers, Gamlin's system penalizes inaccuracy and therefore discourages risk-taking. On that latter note, the left side of the board is sometimes called the married man's side, as it is the better option for those who want to play it safe.

▼

MOVING to winter sports in North America, the National Basketball Association and the National Hockey League both have regular seasons consisting of 82 games.

83 [prime]

WHAT property unites the following 83 five-digit numbers? (Commas within the numbers are deleted so as not to drive you batty.) (See Answers.)

0
11826, 12363, 12543, 14676, 15681, 15963, 18072, 19023, 19377, 19569, 19629, 20316, 22887, 23019, 23178, 23439, 24237, 24276, 24441, 24807, 25059, 25572, 25941, 26409, 26733, 27129, 27273, 29034, 29106, 30384

2
12586, 13343, 14098, 17816, 21397, 21901, 23728, 28256, 28346

5
10136, 13147, 13268, 16549, 20513, 21877, 25279, 26152, 27209, 28582

8
10124, 10214, 14743, 15353, 17252, 20089, 21439, 22175, 22456, 23113, 26351, 28171

9
10128, 10278, 12582, 13278, 13434, 13545, 13698, 14442, 14766, 16854, 17529, 17778, 20754, 21744, 21801, 23682, 23889, 24009, 27105, 27984, 28731, 29208

There is a huge hint to this puzzle elsewhere in this book. For now, though, your only hint—other than the left-hand marking numbers, whose meaning you'll have to figure out—is that the only numbers that could conceivably have this property must be between 10000 and 31622. But the numbers above are the only 83 that actually work.

▼

IF you list all the positive integers from 1 to 500,000,000, how many 1's appear altogether? And why is this question being asked here, on the page for the number 83? (See Answers.)

84 $\left[\, 2^2 \times 3 \times 7 \,\right]$

ONE of history's oldest and best-known algebra problems goes by the name of Diophantus's Riddle, and it reads as follows:

"Here lies Diophantus," the wonder behold. Through art algebraic, the stone tells how old: "God gave him his boyhood one-sixth of his life, One-twelfth more as youth while whiskers grew rife; And then yet one-seventh ere marriage begun; In five years there came a bouncing new son. Alas, the dear child of master and sage, after attaining half the measure of his father's life chill fate took him. After consoling his fate by the science of numbers for four years, he ended his life."

How old was Diophantus when he died? The solution is readily obtained by converting the text into a simple algebraic equation with a single variable. If you let x = Diophantus's age, then, according to the text, $\frac{x}{6} + \frac{x}{12} + \frac{x}{7} + 5 + \frac{x}{2} + 4 = x$.

Note that the least common denominator of 6, 12, and 7 equals 84. Reworking the equation yields $14x + 7x + 12x + 42x + (84 \times 9) = 84x$, so $75x + (84 \times 9) = 84x$, therefore $84 \times 9 = 9x$ and therefore $x = 84$.

History did record that Diophantus had a son who died at age 42.

▼

ON the subject of old mathematical puzzles, the Rhind Papyrus of the Egyptian Middle Kingdom (circa 1650 BC) contained 84 mathematics problems of various sorts. Rhind was actually a Scotsman, not an Egyptian: He purchased the papyrus in 1858 in Luxor, Egypt, following its discovery there in what may have been illegal excavations. The papyrus measures about a foot wide and 18 feet long. It also goes by the name Ahmes Papyrus, in honor of the scribe who copied it down way back when.

In one of the problems, Ahmes appears to equate the area of a circular field whose diameter is nine units with the area of a square having a side of eight units. Such an inequality would imply that the ratio of the circumference of that circle to its diameter—a ratio we recognize immediately as the definition of π—would be $3\frac{1}{6}$, not exactly right but perhaps not bad by 1650 BC standards.

85 $\left[5 \times 17 \right]$

"A well-tied tie is the first serious step in life."
—Oscar Wilde, c. 1880, *A Woman of No Importance*

IN Oscar Wilde's day, the traditional four-in-hand knot (shown above) was really the only thing going, at least for conventional neckties. The full and half Windsor knots didn't come until the 1930s, and the Pratt knot, invented by Jerry Pratt and popularized by news anchorman Don Shelby, took another fifty years to emerge. But necktie history was accelerated in 1999, when Cambridge University's Thomas Fink and Yong Mao determined through exhaustive study that there are precisely 85 ways of tying a necktie. Most of these 85 knots look absolutely terrible, but in addition to the four known knots, the two men uncovered another six knots deemed elegant enough for actual use.

Fink and Mao's methodology incorporated a basic set of six moves: R_I, R_O, C_I, C_O, L_I, and L_O, the letters denoting left to right, center, and right to left, and the subscripts denoting "out of the shirt" or "into the shirt." The ending move, or T, meant "through," as in through whatever loop had been created by the previous maneuvers. For any given number of three or more "half winds," denoted by an h, Fink and Mao computed the associated number of knots, $K(h)$, as equaling the expression $(\frac{1}{3})(2^{h-2} - (-1)^{h-2})$. The finite length of a tie, coupled with basic aesthetic principles, led Fink and Mao to limit the number of "half winds" to 9, so the total number of knots became:

$$K = \sum K(i) = 1 + 1 + 3 + 5 + 11 + 21 + 43 = 85$$

In other words, while the choice of 9 as a cutoff point had a small element of arbitrariness to it, the final result did not.

Knot sequences can be represented as random walks on a triangular/hexagonal lattice as depicted below, with the directions L, R, and C denoted by the arrows.

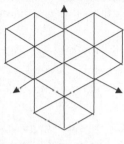

ON some old American cars, the speedometer did not exceed 85 miles per hour. Today, although speedometers uniformly allow for higher rates of speed, the posted speed limits do not.

Of course, lower speed limits are associated with fuel savings, and in that respect 85 mph does not qualify. But in certain areas of the United States, the number 85 does align itself with fuel savings, in the form of E-85, a mixture of 85% ethanol and 15% unleaded gasoline (the gasoline component promotes easier starting in cold weather, among other things). Most cars don't accept this mixture, although the development of E-85 fleets began in 1992.

ETHANOL derives from corn, and the number 85 makes a final appearance in the direct payment formula for farmers of corn and other commodities, the underlying assumption being that farmers use 85% of their "base" acreage:

$$DP_{corn} = (Payment\ rate)_{corn} \times (Payment\ yield)_{corn} \times [(Base\ acres)_{corn} \times 0.85]$$

86 [2 × 43]

TO "eighty-six" something means to get rid of it. How, you may ask, did this usage come to be? The list of possible explanations sounds like a takeoff of *What's My Line?* or some other TV guessing game show. The following explanations suggest that the expression originated in New York City:

A. At New York's famous Delmonico's restaurant, the house steak was number 86 on the menu. Because the restaurant frequently ran out of this item, "86" came to mean something that was no longer available.

B. Article 86 of the New York state liquor code defined the conditions under which a customer would not be served alcoholic beverages.

C. Chumley's, a well-known New York speakeasy, was located at 86 Bedford Street.

D. The elevators at the Empire State Building stop at the eighty-sixth floor.

Alternative explanations include the observation that missing soldiers came to be known as "86'ed" because being AWOL was a violation of Subchapter X Article 86 of the Uniform Code of Military Justice. Finally, there is the theory that the expression is simply a variation of the phrase "deep six." Unfortunately, this last explanation, though the least colorful, is the choice of most etymologists. We close the discussion by noting that Agent 86, otherwise known as Maxwell Smart of *Get Smart* fame, apparently got his number for its suggestion of expendability. But as long as the public maintains its thirst for bogus explanations, I would point out that *Danger Man*, starring Patrick McGoohan (known as *Secret Agent* in the United States), ran for precisely 86 episodes and was created by a gentleman named Ralph Smart.

$2^{86} = 77,371,252,455,336,267,181,195,264$, a number with no zeroes. No larger power of 2 is known to be zero-free.

87 [3 × 29]

THE number 87 is spoken as *quatre-vingts sept* in French—literally "four twenties and a seven." This construction was made famous in America by Abraham Lincoln's Gettysburg Address, which began "Fourscore and seven years ago," a recognition of the 87 years between the signing of the Declaration of Independence in 1776 and the Battle of Gettysburg in 1863.

▼

THE word *decimoctoseptology* won't be found in any dictionary, but it means the study of the number 87, at least to a handful of practitioners who share the quirky view that 87 is the most random number.

▼

IN Australian cricket, a score of 87 is considered unlucky. The superstition supposedly began in 1929 when superstar-to-be Keith Miller went to the Melbourne Cricket Ground to watch legendary batsman Don Bradman bat for New South Wales against Harry "Bull" Alexander of Victoria. Miller, who was but 10 years of age at the time, recalled that Bradman was on 87 when Alexander bowled him. When Miller came of age as a cricketeer, he and South Melbourne teammate Ian Johnson would apparently nudge each other when a batsman or opposing team reached 87, the idea being that an unusual number of batsmen were retired on that number.

Not surprisingly, the data don't support the idea that batsmen fall on 87 any more than on surrounding numbers. Even worse, Miller's memory seems to have failed him, as Bradman was actually on 89 when Alexander bowled him. Yet the reputation of 87 as the "devil's number" endures. The final straw came in 1993, when Bull Alexander died . . . at age 87.

▼

CERTAINLY 87 isn't considered an unlucky number in hockey. Legendary Canadian junior player Sidney Crosby chose 87 as his uniform number because of his birthdate (8/7/87), and continued to wear the number upon entering the NHL in 2005.

▼

THERE are 87 numbers whose squares are a rearrangement of the ten digits 0, 1, 2, 3, 4, 5, 6, 7, 8, and 9 ($32043^2 = 1026753849$, and so on):

32043, 32286, 33144, 35172, 39147, 45624, 55446, 68763, 83919, 99066
35337, 35757, 35853, 37176, 37905, 38772, 39336, 40545, 42744, 43902,
44016, 45567, 46587, 48852, 49314, 49353, 50706, 53976, 54918, 55524,
55581, 55626, 56532, 57321, 58413, 58455, 58554, 59403, 60984, 61575,
61866, 62679, 62961, 63051, 63129, 65634, 65637, 66105, 66276, 67677,
68781, 69513, 71433, 72621, 75759, 76047, 76182, 77346, 78072, 78453,
80361, 80445, 81222, 81945, 84648, 85353, 85743, 85803, 86073, 87639,
88623, 89079, 89145, 89355, 89523, 90144, 90153, 90198, 91248, 91605,
92214, 94695, 95154, 96702, 97779, 98055, 98802

88 $\left[\, 2^3 \times 11 \,\right]$

A piano has 88 keys. There being 7 white keys and 5 black keys to an octave, the full keyboard consists of slightly more than 7 octaves.

▼

AN n-digit number is called "narcissistic" if the sum of the nth powers of its digits equals the number itself. All one-digit numbers are narcissistic by definition, but narcissism gets increasingly rare as the number of digits goes up. Altogether there are 88 narcissistic numbers, the largest being the 39-digit monstrosity 115,132,219,018,763,992,565,095,597,973,971,522,401.

That's right: $1^{39} + 1^{39} + 5^{39} + 1^{39} + 3^{39} + 2^{39} + 2^{39} + 1^{39} + 9^{39} + 0^{39} + 1^{39} + 8^{39} + 7^{39} + 6^{39} + 3^{39} + 9^{39} + 9^{39} + 2^{39} + 5^{39} + 6^{39} + 5^{39} + 0^{39} + 9^{39} + 5^{39} + 5^{39} + 9^{39} + 7^{39} + 9^{39} + 7^{39} + 3^{39} + 9^{39} + 7^{39} + 1^{39} + 5^{39} + 2^{39} + 2^{39} + 4^{39} + 0^{39} + 1^{39} = 115,132,219,018,763,992,565,095,597,973,971,522,401.$ (See **153**.)

THE number 88 reads as *ba ba* in Chinese and has come to mean "so long" in Chinese Internet shorthand.

THE bingo call for 88 is "two fat ladies." UK residents will recognize the term as the title of a late 1990s TV show starring Clarissa Dickson Wright and the late Jennifer Paterson. The two stars drove around the countryside on a Triumph Thunderbird with a sidecar: the motorcycle's plate number was N88TFL.

IF the two fat ladies happened to be driving at 60 miles per hour, they'd be going 88 feet per second, according to a standard conversion formula:

60 miles/hour $\times$ 5280 feet/mile $\div$ 3600 seconds/hour = 88 feet/second

And even 88 *miles per hour* has some historical significance, it being the speed at which Michael J. Fox's DeLorean would enter into time-travel mode in the *Back to the Future* trilogy.

88 is the fourth "untouchable" number, by which it is meant a number that is not the sum of the proper divisors of any other number. (The first three untouchable numbers are 2, 5, and 52.) Paul Erdos demonstrated the existence of infinitely many untouchable numbers, but there is only one known odd untouchable number, namely 5. Are there others? Well, oddly enough, this question is tied to Goldbach's Conjecture, one of the most famous unsolved problems in number theory. Goldbach's Conjecture, first suggested by Prussian mathematician Christian Goldbach in 1742 and still unproven at this writing, is the simple-looking assertion that any even number is the sum of two primes. Suppose for a moment that this conjecture is true, and look at the odd number $2n + 1$. According to Goldbach's conjecture, we may write $2n = p + q$ for some primes p and q. But the sum of the proper factors of the number pq is therefore $1 + p + q = 2n + 1$, so the original odd number $2n + 1$ can't be untouchable.

89 [prime $8^1 + 9^2$]

89 is the only two-digit number that can be expressed as the sum of its digits raised to the consecutive powers 1 and 2—that is, the position of the digits from left to right. (See **135** and **175**.)

▼

89 is the eleventh Fibonacci number, and the reciprocal of 89 bears a curious relationship with the Fibonacci sequence. Create a triangle of numbers such that the rightmost digit of the *n*th Fibonacci number is in the *n+1st* decimal place.

.01
.001
.0002
.00003
.000005
.0000008
.00000013
.000000021
.0000000034
etc.

_____

The sum of these numbers = .01123595505618 . . . = $\frac{1}{89}$

Although this result is surprising, the proof is relatively straightforward. The idea is to let *x* equal the sum in question, and then use the fundamental Fibonacci relationship ($F_{n+1} = F_n + F_{n-1}$) to produce the equation $100x - 10x - x = 1$. Because the left side equals $89x$, you get $89x = 1$, or $x = \frac{1}{89}$. What makes 89 special in this equation is not so much that it is a Fibonacci number, but that it equals $100 - 10 - 1$.

▼

THE appearance of Fibonacci numbers in nature is well-known if not always exact. It seems that sunflowers often have 55 (F_{10}) clockwise spirals and 89 (F_{11}) counterclockwise spirals.

89 is a Sophie Germain prime, meaning that $2 \times 89 + 1$ is also prime. As it happens, starting with 89 and continuing in this fashion creates a sequence of six primes, given below. (The longest-known such chain contains 16 primes, starting with 810,433,818,265,726,529,159.)

89	2A + 1	2B + 1	2C + 1	2D + 1	2E + 1
89	179	359	719	1439	2879
A	B	C	D	E	F

In 1825, Sophie Germain proved that there is no solution to the equation $x^p + y^p = z^p$ if p is a Sophie Germain prime, one small step on the road to proving Fermat's Last Theorem.

THE definition of Sophie Germain primes and the magnitude of the largest known such prime were mentioned by the characters Hal and Catherine in the 2005 film *Proof*. Not long after the film came out (May 2006), an even larger Sophie Germain prime was discovered. It has 51,780 digits, a bit much for this page, but we can describe it more compactly as $p = 137211941292195 \times 2^{171960} - 1$.

90 $\left[\ 2 \times 3^2 \times 5 \qquad 9^1 + 9^2 \qquad (15 - 9) \times (15 - 0)\ \right]$

AN angle whose measure is 90 degrees is called a right angle. In radian measure, 90 degrees corresponds to $\frac{\pi}{2}$ radians.

A baseball diamond not only has four right angles; it has 90 feet between bases. Said columnist Red Smith (1905–1982), "Ninety feet between home and first base is perhaps as close as man has ever come to perfection."

THE pattern below is a sample of dried-out mud. Although the pattern contains much curvature, the equilibrating pressures at intersecting lines create angles very close to 90 degrees.

IN *Gulliver's Travels*, the people of Laputa evidently lacked instruments such as the T-square, whose purpose is to create 90-degree angles. Wrote Swift, "Their houses are very ill built, the walls bevil, without one right angle in any apartment."

▼

THE number of primes less than 90 equals the number of integers less than 90 that are *relatively* prime to 90 (i.e., share no common factor with 90). There are only 7 numbers with this same property, and 90 is the largest of those seven. (The others are 2, 3, 4, 8, 14, and 20.) So, for example, there are six primes less than 14 (2, 3, 5, 7, 11, and 13) and six numbers less than 14 that are relatively prime to 14 (1, 3, 5, 9, 11, and 13).

91 $\left[\, 7 \times 13 \,\right]$

IN the card game Diamond Points, the diamond suit is separated from the rest of a deck of cards and is placed facedown in a stack. Players (two or three, ideally) are each given another suit. The diamonds are revealed one at a time and players vie for the points represented by the diamond (Ace = 1 . . . King = 13) by selecting a card from their given suit, the high card winning the "diamond points." In a two-player game, it is sufficient to win 46 points, because the total number of diamond points equals $1 + 2 + \ldots + 13 = 91$. You will perhaps recognize this last equation as establishing 91 as the thirteenth triangular number.

▼

THE equation $91 = 1^2 + 2^2 + 3^2 + 4^2 + 5^2 + 6^2$ shows that 91 is the sum of the first six squares, so 91 is a square-pyramidal number, and is the first triangular and square-pyramidal number that we have encountered since 55. But that gap is stunningly small for such a mathematical rarity. The next number that is both triangular and square-pyramidal—208,335—is also the *last* number with both properties.

BY using negative numbers, we can write 91 as the sum of two cubes in two different ways, namely $91 = 4^3 + 3^3 = 6^3 + (-5)^3$.

We can also write $91 = 1 + 5 + 10 + 25 + 50$, so if you had every US coin short of a silver dollar, you'd have a total of 91 cents. Elsewhere in the world of US finance, the term "91-day T-bill" arises because 91 days represents one-quarter of a year. Bills that mature in 91 days are the shortest maturity securities issued by the US Treasury.

THE multiplication table for 91 yields a cute result. If you look at the three columns individually, they speak for themselves:

91	×	1	=	9 1
		2	=	1 8 2
		3	=	2 7 3
		4	=	3 6 4
		5	=	4 5 5
		6	=	5 4 6
		7	=	6 3 7
		8	=	7 2 8
		9	=	8 1 9

ONE classic version of the abacus consists of 13 columns, each with 7 disks, for a total of 91 disks. The Babylonians are credited with the earliest form of the abacus, and calculations using the abacus are referenced in the writings of Greek scholars such as Herodotus and  Demosthenes. The separation of the abacus into two zones is a Japanese innovation, and the instrument is of course associated with the Far East to this day. Perhaps the greatest day in abacus history came in 1946, when abacus-wielding Kiyoshi Matsuzaki won a speed calculation contest against a US Army Private (T. N. Wood) equipped with a state-of-the-art mechanical calculator.

The worst day in abacus history, probably during the same era, came when an abacus salesman entered a restaurant in Brazil and challenged a customer to a calculating contest. Although the salesman proved to be the speedier one at addition and multiplication, he faltered upon upping the ante to *raios cubicos*, or cube roots. His specific undoing was in choosing 1729.03, a number that the customer recognized as just fractionally higher than 1728, or 12 cubed. Within seconds, the customer jotted down 12.002 as the cube root of 1729.03, an estimate that left the abacus salesman in the dust. The salesman left in disgrace, presumably never knowing that the man in the restaurant was none other than Richard Feynman, a Nobel laureate-to-be and one of the most sparkling minds of his generation.

92 $\left[\ 2^2 \times 23\ \right]$

THERE are 92 ways of placing 8 queens on an 8 × 8 chessboard such that no queen is under attack from another. Below are the 12 basic solutions, which grow to 92 by suitable rotations and reflections (as long as you don't mind having a black square in the lower right-hand corner, contrary to the orientation of the board in an actual chess game):

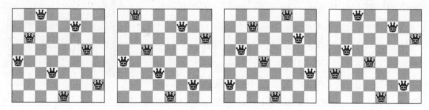

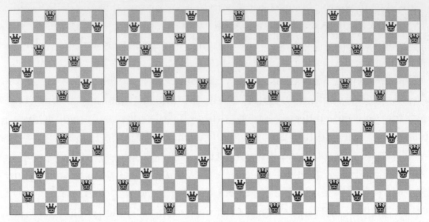

Let's think about that last sentence for just a minute. Exactly how do you go from 12 to 92? It's not obvious, is it? If you start with, say, the top-left image, you can rotate it in any of four ways (three 90-degree turns and another to return to the starting position) and you can reflect it in two ways (upside down or right-side up). Altogether that's $4 \times 2 = 8$ solutions out of that one image. If you followed the same procedure with each of the 12 images, you'd expect to generate $8 \times 12 = 96$ total solutions. But in real life you only get to 92, a number that isn't even divisible by 12. What gives?

The answer is that one of the 12 basic solutions doesn't quite carry its own weight. The rogue is the very last one, at the bottom right. Because the queens in that solution are placed symmetrically with respect to the center of the board, rotations and reflections only produce 4 solutions, not 8, so the total number is $8 \times 11 + 4 = 92$.

The 8-queens puzzle was introduced in 1848 by a chess player named Max Bezzel. The problem was solved in 1850 by Franz Nauck, who extended the problem to queens on an $n \times n$ chessboard, where n can take any value. For example, on a 24×24 board, the total number of solutions equals 2,275,141,71,973,736. It is unreasonable to ask that you produce all of these solutions, so here's an easier, somewhat related problem: What is the smallest number of queens that can be placed on an 8×8 chessboard so that every square is attacked by at least one of the queens? (See Answers.)

93 $[\,3 \times 31\,]$

THE first time a schoolchild is apt to encounter the number 93 is upon learning that the sun is, on average, 93 million miles from Earth. That distance is officially known as an astronomical unit. The defrocked planet Pluto, for example, is 39.5 AUs from the sun, plus or minus 9.8 depending on Pluto's position along its elliptical orbit. Progress in astronomy being what it was, the AU was in use as a relative measure long before scientists knew what its exact distance was.

▼

QUATREVINGT-TREIZE was Victor Hugo's final novel. The reference of the book's title is to 1793, the most horrific year of the French Revolution, and in particular the year in which Marie Antoinette of "Let them eat cake" fame was sent to the guillotine.

▼

OF course, had the French put the guillotine to a more benign use they would have discovered that it is possible to cut a cake into 93 pieces using only 8 straight cuts.

And we can generate that number by taking advantage of a tidy mathematical relationship between cutting in three dimensions versus just two.

Recall (see **22**) that in two dimensions the number of pieces of pizza that can be generated from n straight cuts equals $\frac{(n^2 + n + 2)}{2}$, or just 1 more than the nth triangular number. (The discussion there involved drawing straight lines, as opposed to making cuts, but the two approaches are equivalent.) Here's how the two-dimensional sequence begins:

# of cuts (n):	0	1	2	3	4	5	6	7	8	9	10
		+	+	+	+	+	+	+	+	+	+
Max # pieces from n cuts in 2 dimensions	1	2	4	7	11	16	22	29	37	46	56

To explain the plus signs, once you place a 1 at beginning of the bottom row, each of the following numbers in that row can be obtained by adding the number to its left to the number above it—in other words, you add along the diagonals and then drop. Let's try that same procedure, replacing the top row with the bottom row (except that we add a 0 in the first position) and again starting with a solitary 1 on the new bottom row:

0	1	2	4	7	11	16	22	29
	+	+	+	+	+	+	+	+
1								

In short order the table fills out as below.

# of cuts (n):	0	1	2	3	4	5	6	7	8
2-dim seq for (n-1)	0	1	2	4	7	11	16	22	29
		+	+	+	+	+	+	+	+
Max # of pieces from n cuts in 3 dimensions	1	2	4	8	15	26	42	64	93

Remarkably, the bottom row now consists of the maximum number of pieces that can be generated by n cuts in *three* dimensions, culminating with 93 pieces from 8 cuts. (Pizza is considered two-dimensional in this cutting exercise, whereas a cake represents a three-dimensional object.)

What makes the following 93 numbers special? And what is different about the ones in boldface? (See Answers.)

10301 10501 10601 11311 **11411 12421 12721 12821 13331** 13831 13931 14341 **14741**
15451 15551 16061 16361 16561 **16661** 17471 17971 18181 18481 19391 **19891** 19991
30103 30203 30403 **30703 30803 31013 31513 32323 32423** 33533 34543 34843
35053 35153 35353 35753 **36263** 36563 37273 37573 **38083 38183** 38783 39293
70207 **70507 70607** 71317 71917 **72227 72727** 73037 73237 73637 **74047 74747**
75557 **76367 76667 77377 77477 77977** 78487 78787 78887 79397 **79697 79997**

90709 91019 93139 **93239** 93739 94049 94349 **94649** 94849 **94949**
95959 96269 96469 **96769** 97379 97579 97879 98389 98689

The only hint you'll get is that these two questions, although juxtaposed, are on completely different wavelengths.

94 [2 × 47]

THE factorization of 94 tells us at a glance that it is not divisible by 4. When the Winter Olympics were held in Lillehammer, Norway, in 1994, it marked the first time that the Modern Olympics, either summer or winter, were held on a year not divisible by 4. The idea introduced at that time was to stagger the winter and summer games, creating Olympic action every two years rather than a whole lot of action every four years. From this point forward, the Olympic Games will operate on two separate four-year cycles, with the summer games taking years divisible by 4 and the winter games the remaining even years: as a mathematician would say, those years that are congruent to 2, mod 4.

▼

BOTH digits in 94 are perfect squares, as demonstrated by the diagram, which has nine dots altogether, four of which are white. But this diagram is the beginning of a very different route to the number 94. On the next page we see that there are plenty of other ways to choose four points so as to form a quadrilateral, from the trapezoid on the left to the parallelogram to the kite (yes, that's what the third figure is called) and finally to the nameless figure on the far right (and plenty more).

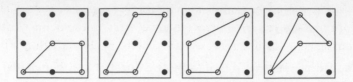

How many ways can four points be chosen so that their vertices form a quadrilateral? Altogether it is possible to form 6 squares, 4 rectangles, 12 parallelograms, 28 trapezoids, 8 kites, 16 other convex shapes, and 24 non-convex shapes, for a total of 94.

▼

CHECK out the measurements of the following old ships:

Ship	Launched	Tonnage	Crew
HMS *Speedy*	1782	208 8/94	90
HMS *Dart*	1796	386 16/94	140
HMS *Lively*	1804	1071 90/94	284
HMS *Surprise*	1794	578 73/94	200
HMS *Boadicea*	1797	1052 5/94	282

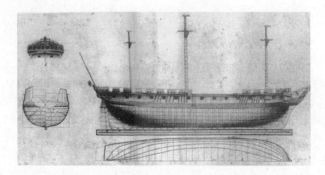

What's with the 94's in the denominators? They arise from the formula for Builder's Old Measurement, the prevailing standard for a ship's cargo-carrying capacity (tonnage) until the introduction of steam propulsion in the mid-nineteenth century. The precise formula is given by $T = (L - 3B/5)B^2/2/94$, where T = tonnage, L = length, and B = breadth. (Thames

Measurement, a predecessor formula for a ship's capacity, has a slightly different formula but was also characterized by a 94 in the denominator.)

Of course, all this still doesn't explain where the 94 in the formula comes from. Its origin turns out to depend on an old standard declaring that a ship's tax burden should be $\frac{3}{5}$ of its displacement. The formula for displacement is Length × Beam × Draft × Block Coefficient, all divided by 35 cubic feet per ton of seawater. The Beam is defined as the ship's widest point, while the Draft (the distance between the bottom of the ship and the water line, is estimated to be half the Beam. The Block Coefficient is estimated at 0.62; if you draw a box around the submerged portion of the ship, the block coefficient is the fraction of that box represented by the ship's volume. Multiplying $\frac{3}{5}$ by .62 and dividing by 35 gets us awfully close to $\frac{1}{94}$, hence the appearance of that fraction in the final formula.

I know. That was a lot of work just to track down a weird denominator, but there you have it.

95 $\left[\, 5 \times 19 \,\right]$

THE number 95 plays a time-honored role with respect to survey data and confidence intervals. When a poll is released indicating that, say, 57% of the respondents favor a certain political candidate, underlying this calculation is a margin of error and a confidence interval. If the poll has a margin of error of three percentage points and a confidence interval of 85%, that means that if the poll were to be conducted 100 times, you'd expect the percentage of respondents favoring that candidate would be between 54% and 60% on 85 occasions—three percentage points to either side of the announced figure. This delineation is not well understood, in part because most polls release a margin of error but not a confidence interval. The reason for this omission is that it's not really an omission at all. The vast majority of polls use a confidence interval of 95%.

▼

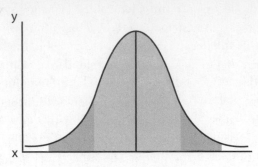

WHEN data follow a normal distribution (the bell curve shown above), the number 95 makes a more specific appearance. One nice thing about such distributions is that their means and standard deviations are either known or readily calculated. It turns out that when data follow a bell curve, 95% of all observations are within two standard deviations of the mean, as marked by the thick vertical bands to either side of the peak in the curve above.

▼

PERHAPS the most historically significant appearance of 95 came in 1517, when Martin Luther nailed his 95 Theses to the door of the Castle Church in Wittenberg, Germany. Space considerations don't permit a full listing of Luther's theses, so here is number 16, chosen for its relative lack of controversy: "Hell, purgatory, and heaven seem to differ as do despair, almost-despair, and the assurance of safety."

▼

IN computing, the list of displayable ASCII (American Standard Code for Information Interchange) characters comprises 26 uppercase letters, 26 lowercase letters, 10 digits, and 33 special characters, including punctuation. That's a grand total of 95.

96 $\left[\ 2^5 \times 3\ \right]$

FORMER major-league baseball player Bill Voiselle, a onetime pitcher for the New York Giants, Boston Braves, and Chicago Cubs, got special dispensation from the National League to wear the number 96, at the time the highest number ever worn by a major leaguer. Why? Because he grew up in the town of Ninety Six, South Carolina. The town was so named because it was thought to be 96 miles away from the Cherokee settlement Keowee (even though it really wasn't). Apparently a bill to change the name of Ninety Six to Cambridge once appeared before the state legislature, but a Ninety Six resident held up a sign with the number 96 on it, pointing out that it read the same right-side up as upside down, and so it should remain. And so it has.

▼

$$96 = 2 + 5 + 7 + 10 + 30 + 42 \text{ and } \frac{1}{2} + \frac{1}{5} + \frac{1}{7} + \frac{1}{10} + \frac{1}{30} + \frac{1}{42} = 1$$

$$96 = 6 + 7 + 7 + 8 + 8 + 9 + 12 + 18 + 21 \text{ and } \frac{1}{6} + \frac{1}{7} + \frac{1}{7} + \frac{1}{8} + \frac{1}{8} + \frac{1}{9} + \frac{1}{12} + \frac{1}{18} + \frac{1}{21} = 1$$

ABOVE are two partitions of 96 whose reciprocals add to 1. Such a partition is called "exact." Believe it or not, there are precisely 96 exact partitions of 96.

▼

THE game of Ishido is played on a board measuring 8 squares × 12 squares, for 96 squares altogether. Although it has the look of an ancient game, it was introduced in 1990.

97 $\left[\ \text{prime}\ \right]$

THE Gregorian calendar has a cycle of 400 years, during which time there are 97 leap years. In theory, you'd expect one leap year every four years, for a total of 100, but only one of the four "century years" in a 400-year span (the one that's divisible by 400) is a leap year.

▼

$\frac{1}{97}$ = 0.01030927 . . . Note that if you multiply the first pair of digits after the decimal point by 3, you get the second pair, and so on until you get to 27. No, this pattern doesn't continue, but it was nice while it lasted.

▼

WHEREAS a standard tarot deck consists of 78 cards, the Minchiate Tarot deck of the Renaissance contained 97. The extra cards were four Virtues (Prudence, Hope, Faith, and Charity), the four elements (Earth, Air, Fire, and Water), and the 12 signs of the zodiac. That's 20 extra cards, but for some reason the High Priestess of the standard tarot deck was not included in the Minchiate deck, so the final total was 97.

▼

WHEN written out, NINETY-SEVEN alternates consonants and vowels, and is the longest number to do so. Sort of. Can you come up with a longer one? (See Answers.)

▼

THERE are 97 ways of using the 10 digits 0 through 9 to form two fractions that add up to 1. On the assumption that you might not believe me, here they are:

$\frac{3485}{6970} + \frac{1}{2}$	$\frac{3548}{7096} + \frac{1}{2}$	$\frac{3845}{7690} + \frac{1}{2}$	$\frac{4538}{9076} + \frac{1}{2}$
$\frac{4685}{9370} + \frac{1}{2}$	$\frac{4835}{9670} + \frac{1}{2}$	$\frac{4853}{9706} + \frac{1}{2}$	$\frac{4865}{9730} + \frac{1}{2}$
$\frac{7365}{9820} + \frac{1}{4}$	$\frac{3079}{6158} + \frac{2}{4}$	$\frac{1278}{6390} + \frac{4}{5}$	$\frac{1872}{9360} + \frac{4}{5}$
$\frac{7835}{9402} + \frac{1}{6}$	$\frac{3190}{4785} + \frac{2}{6}$	$\frac{1485}{2970} + \frac{3}{6}$	$\frac{2079}{4158} + \frac{3}{6}$
$\frac{2709}{5418} + \frac{3}{6}$	$\frac{2907}{5814} + \frac{3}{6}$	$\frac{4851}{9702} + \frac{3}{6}$	$\frac{4362}{5089} + \frac{1}{7}$
$\frac{5940}{8316} + \frac{2}{7}$	$\frac{6810}{9534} + \frac{2}{7}$	$\frac{5803}{7461} + \frac{2}{9}$	$\frac{1208}{5436} + \frac{7}{9}$

$$\frac{1352}{6084} + \frac{7}{9} \qquad \frac{729}{3645} + \frac{8}{10} \qquad \frac{927}{4635} + \frac{8}{10} \qquad \frac{876}{3504} + \frac{9}{12}$$

$$\frac{485}{970} + \frac{13}{26} \qquad \frac{369}{574} + \frac{10}{28} \qquad \frac{486}{972} + \frac{15}{30} \qquad \frac{485}{970} + \frac{16}{32}$$

$$\frac{287}{369} + \frac{10}{45} \qquad \frac{728}{936} + \frac{10}{45} \qquad \frac{169}{507} + \frac{32}{48} \qquad \frac{269}{807} + \frac{34}{51}$$

$$\frac{204}{867} + \frac{39}{51} \qquad \frac{678}{904} + \frac{13}{52} \qquad \frac{893}{1026} + \frac{7}{54} \qquad \frac{609}{783} + \frac{12}{54}$$

$$\frac{309}{618} + \frac{27}{54} \qquad \frac{308}{462} + \frac{19}{57} \qquad \frac{273}{406} + \frac{19}{58} \qquad \frac{307}{614} + \frac{29}{58}$$

$$\frac{748}{935} + \frac{12}{60} \qquad \frac{207}{549} + \frac{38}{61} \qquad \frac{208}{793} + \frac{45}{61} \qquad \frac{485}{970} + \frac{31}{62}$$

$$\frac{507}{819} + \frac{24}{63} \qquad \frac{284}{710} + \frac{39}{65} \qquad \frac{148}{296} + \frac{35}{70} \qquad \frac{481}{962} + \frac{35}{70}$$

$$\frac{145}{290} + \frac{38}{76} \qquad \frac{451}{902} + \frac{38}{76} \qquad \frac{417}{695} + \frac{32}{80} \qquad \frac{306}{459} + \frac{27}{81}$$

$$\frac{630}{945} + \frac{27}{81} \qquad \frac{405}{729} + \frac{36}{81} \qquad \frac{540}{972} + \frac{36}{81} \qquad \frac{60}{1245} + \frac{79}{03}$$

$$\frac{109}{327} + \frac{56}{84} \qquad \frac{307}{921} + \frac{56}{84} \qquad \frac{310}{465} + \frac{29}{87} \qquad \frac{315}{609} + \frac{42}{87}$$

$$\frac{231}{609} + \frac{54}{87} \qquad \frac{504}{623} + \frac{17}{89} \qquad \frac{105}{623} + \frac{74}{89} \qquad \frac{276}{345} + \frac{18}{90}$$

$$\frac{372}{465} + \frac{18}{90} \qquad \frac{138}{276} + \frac{45}{90} \qquad \frac{186}{372} + \frac{45}{90} \qquad \frac{381}{762} + \frac{45}{90}$$

$$\frac{185}{370} + \frac{46}{92} \qquad \frac{140}{368} + \frac{57}{92} \qquad \frac{426}{710} + \frac{38}{95} \qquad \frac{473}{528} + \frac{10}{96}$$

$$\frac{357}{408} + \frac{12}{96} \qquad \frac{735}{840} + \frac{12}{96} \qquad \frac{375}{480} + \frac{21}{96} \qquad \frac{531}{708} + \frac{24}{96}$$

$$\frac{135}{270} + \frac{48}{96} \qquad \frac{351}{702} + \frac{48}{96} \qquad \frac{143}{528} + \frac{70}{96} \qquad \frac{34}{578} + \frac{96}{102}$$

$$\frac{693}{728} + \frac{5}{104} \qquad \frac{59}{236} + \frac{78}{104} \qquad \frac{63}{728} + \frac{95}{104} \qquad \frac{56}{832} + \frac{97}{104}$$

$$\frac{56}{428} + \frac{93}{107} \qquad \frac{87}{435} + \frac{96}{120} \qquad \frac{496}{508} + \frac{3}{127} \qquad \frac{67}{204} + \frac{98}{136}$$

$$\frac{795}{810} + \frac{6}{324} \qquad \frac{684}{702} + \frac{9}{351} \qquad \frac{792}{801} + \frac{4}{356} \qquad \frac{693}{704} + \frac{8}{512}$$

$$\frac{792}{801} + \frac{6}{534}$$

98 $\left[\, 2 \times 7^2 \,\right]$

THE number 98 makes a pivotal appearance in the "potato paradox." Specifically, suppose you start with 100 pounds of potatoes, which you understand to be 99% water. Over time, as the water in the potatoes evaporates, this percentage figure decreases. By the time the potatoes are 98% water, how much do they weigh? (See Answers.)

▼

PICTURED is one of 98 theoretical tic-tac-toe (noughts and crosses) patterns in which **X** has a win. However, not all of these positions are realizable in an actual game, because in some of those patterns **O** would also have a win. (See **62**.) Of course, any game of tic-tac-toe that is not a draw is pretty suspect to begin with, so you're more likely to encounter the above outcome in a game of *random* tic-tac-toe.

X	X	X
O	O	X
O	X	O

▼

AS of the year 2000, 98 is the highest number that can be worn by players in the National Hockey League. Of course, this has less to do with the number 98 than with the fact that (1) no player can wear a number with more than two digits, and (2) 2000 was the year in which the league permanently retired the 99 worn by Wayne Gretzky from 1978 to 1999. (Gretzky originally wanted Gordie Howe's number 9 back in his junior hockey days, but a teammate already had that number so he settled for 99.)

99 $\left[3^2 \times 11 \right]$

ACCORDING to Thomas Edison, who really should know, genius is 1% inspiration and 99% perspiration.

▼

IF you get a perfect score on a standardized exam, you're still "only" in the 99th percentile. In general, the percentile rank of a test score is the percentage of scores in the overall frequency distribution that are lower, and technically the highest integral value that this number can take on is 99. Of course, there's nothing preventing someone from being in the 99.99 percentile, depending on the nature of the distribution, but percentile ranks on standardized tests don't ordinarily include any decimal points. Even students who score in that rarefied territory should be humbled by another Edison quote: "We don't know a millionth of one percent about anything."

▼

POPULAR culture has seen its fair share of 99's throughout the world. Canada produced the greatest number 99 ever, hockey Hall of Famer Wayne Gretzky; America was home to Barbara Feldon, aka. Agent 99 of *Get Smart* fame; and the German rock group Nena had a smash 1983 hit with "99 Luftballons," a song that made it into English-speaking countries under the name "99 Red Balloons." The song was a Cold War protest in which a bunch of balloons got loose and crossed international borders, triggering a military overreaction.

▼

OF course, the Anglicized version of "99 Luftballons" wasn't the first song to have "99" in the title. That honor surely belongs to "99 Bottles of Beer on the Wall," which in turn derives from the British song "10 Green Bottles." The original version has presumably been sung from start to finish on many occasions, unlike its unwieldy successor.

▼

THE Muslim rosary theoretically has 99 grains, representing the 99 sacred names of Allah. The beads serve as a counting device for incantations in which these names are repeated, but in practice these grains often number 33, the same number as the beads found on Christian rosaries. Because 33 is a factor of 99, this smaller set of beads functions perfectly well as a counting mechanism.

▼

IN the original specifications for compact discs, CD Digital Audio, the CD commonly used in stereo systems, was capable of holding up to 99 tracks.

▼

IN Italian legend, there was once a king who had 99 elite bodyguards, giving the number 99 an association with quality and elegance. In modern legend, "Number 99" ice cream was allegedly so named by Italian expatriates who wanted to convey the notion of high quality. But neither legend holds up very well. Research has revealed that the bodyguards in question must have been the Vatican's Swiss Guard, a group that traditionally numbered 105. And the "99" ice cream—vanilla with a chocolate sliver—apparently was named by the Italian owners of a Scottish ice cream shop not because of any ties to ancient legend, but because the shop was located at 99 High Street. More evidence that you can't believe what you read.

100 $\left[\, 2^2 \times 5^2 \,\right]$

$$100 = (1 + 2 + 3 + 4)^2 \text{ and } 100 = 1 + 8 + 27 + 64 = 1^3 + 2^3 + 3^3 + 4^3$$

IN general, the sum of the first n cubes equals the square of the sum of the first n integers.

▼

IN a base 10 world, it is not surprising that the number 100 shows up in some important places. For example, there are 100 cents in the dollar, the boiling point of water in the Celsius scale, aka centigrade, is 100 degrees, quite by design, and 100 is also the number of US senators, two from each of the 50 states.

▼

THE prefix *cent-* means 100, as in century, centipede, and so forth. In French, *cent* is the word for 100 and is the biggest French number to be in alphabetical order. In fact, look what happens when you spell out the equation $2 \times 5 \times 10 = 100$ in French:

$$\text{DEUX} \times \text{CINQ} \times \text{DIX} = \text{CENT}$$

Each of the numbers is in alphabetical order!

▼

PERHAPS my favorite use of the number 100 comes from China, where tradition holds that the naming of a newborn panda must wait until the cub is 100 days old.

▼

100 equals 10 squared, and there are a couple of 10×10 squares worth noting. This 10×10 word square was sought after for many years. Although imperfect (it uses two obscure place names—Adaletabat and Dioumabana—and a hyphenated expression—nature-name), it is a remarkable accomplishment, perhaps a maximal one; the odds of someone coming along with an 11×11 square are slim indeed.

D	E	S	C	E	N	D	A	N	T
E	C	H	E	N	E	I	D	A	E
S	H	O	R	T	C	O	A	T	S
C	E	R	B	E	R	U	L	U	S
E	N	T	E	R	O	M	E	R	E
N	E	C	R	O	L	A	T	E	R
D	I	O	U	M	A	B	A	N	A
A	D	A	L	E	T	A	B	A	T
N	A	T	U	R	E	N	A	M	E
T	E	S	S	E	R	A	T	E	D

▼

A different type of 10 × 10 square consists of 100 squares and another set of 100 smaller squares nested inside. Although we don't have the luxury of a 10-color presentation, suffice it to say that for both sets of 10 squares, each row and each column consists of 10 different colors.

This square was discovered in 1959 by E. T. Parker of Remington Rand and R. C. Bose and S. Shrikhande of the University of North Carolina. Their work put to rest a longstanding conjecture of Euler, who had posited over a century and three-quarters earlier that such a Graeco-Roman square was impossible if the side of the square was 2, 6, 10, 14 . . . and so on (in general terms, congruent to 2 modulo 4). All other sizes were known to work.

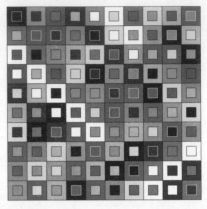

The appearance of the computer company Remington Rand (Univac) on Parker's resume suggests that the discovery was computer-aided, but in fact that wasn't the case. While it would have been possible to harness computer power to create Graeco-Roman squares of certain sizes, the work of Parker, Bose, and Shrikhande was more general, resulting in the proof of the astonishing fact that Graeco-Roman squares were in fact possible for any side lengths whatsoever other than 2 and 6. (See **36**.)

101 [prime]

DODIE Smith's 1956 novel *The Hundred and One Dalmatians* had a plotline that might seem too gruesome for younger audiences, but Walt Disney Productions made the book into a highly successful animated film in 1961 and then a live-action film in 1996.

AT age 20, Joseph Fourier asked if 17 lines could create precisely 101 points of intersection. The diagram to the right represents one of four possible families of solutions. Fourier later revolutionized algebraic and differential equations and even discovered (in 1824) the atmospheric phenomenon that would one day be known as the "greenhouse effect."

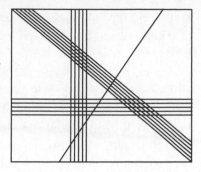

102 [2 × 3 × 17]

THE Empire State Building has a total of 102 stories. Upon its completion in 1931, it surpassed the Chrysler Building and 40 Wall Street to become the world's tallest building. It has its own ZIP code—10118.

103 [prime]

IN the play/movie *Proof*, the lead character is a mathematics professor who upon his death left 103 notebooks, the value of which would be explored by a graduate student in mathematics and his own daughter. The daughter, played by Gwyneth Paltrow in the film version, turned out to be the author of the one piece of groundbreaking mathematics contained in the notebooks.

104 $\left[\ 2^3 \times 13\ \right]$

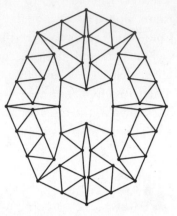

THIS diagram is the creation of German mathematician Heiko Harborth and is the smallest known 4-regular matchstick graph: The 104 matchsticks are arranged so that every vertex in the diagram has four matchsticks emanating from it.

It turns out to be impossible to create an arrangement in which *five* or more matchsticks meet at every vertex. For two matchsticks at each vertex, the answer is an equilateral triangle. Can you find the (12 matchstick) solution in which precisely three meet at each vertex? (See Answers.)

105 $\left[\ 3 \times 5 \times 7\ \right]$

THE smallest number to have three distinct odd prime factors is 105. In advanced mathematics, this fact leads to a surprising result involving $\Phi_{105}(x)$—the so-called cyclotomic polynomial of degree 105. For any positive integer n, cyclotomic polynomials are the building blocks for the expression $x^n - 1$, and they take on some simple forms, as in

$$\Phi_2(x) = x + 1$$
$$\Phi_4(x) = x^2 + 1$$
$$\Phi_7(x) = x^6 + x^5 + x^4 + x^3 + x^2 + x + 1$$

Anyway, it's not much of a punch line, but $\Phi_{105}(x)$ is the first cyclotomic polynomial having any coefficients other than 1 and -1.

106 $\left[\, 2 \times 53 \,\right]$

THE New York Philharmonic, by the numbers:

Violin	33
Viola	12
Cello	11
Bass	9
Flute	4
Piccolo	1
Oboe	2
English horn	1
Clarinet	4
E flat clarinet	1
Bass clarinet	1
Bassoon	4
Contrabassoon	1
Horn	6
Trumpet	3
Trombone	3
Bass trombone	1
Tuba	1
Timpani	1
Percussion	2
Harp	1
Harpsichord	1
Piano	2
Organ	1
TOTAL	**106**

107 [prime]

AM radio signals are assigned within a band between 535 and 1605 kilohertz (kHz). Since each frequency has a bandwidth of 10 kHz, there are $\frac{(1605 - 535)}{10} = 107$ possible carrier frequencies in any given area.

108 [$2^2 \times 3^3$]

PERHAPS the best-known mathematical appearance of 108 comes from the regular pentagon, in which each of the five interior angles measures 108 degrees.

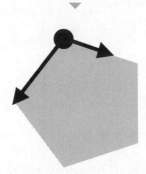

THIS particular pentagon is nested inside a pair of clock hands set at 3:36. Note that the product of the hours and minutes is $3 \times 36 = 108 =$ the number of degrees between the two hands. Can you find the two other times that have this property? (See Answers.)

IN Homer's *Odyssey*, Penelope was wooed by 108 suitors during Odysseus's absence.

THERE are 108 double stitches on an official Major League baseball.

▼

THE game of canasta uses 108 cards—two full decks plus 4 jokers.

109 [prime]

TWICE the sum of the first 109 integers equals 10,900 + 1090.

$\frac{1}{109}$ is a 108-digit repeating decimal that ends with 853211—the beginning of the Fibonacci sequence, only backward. Specifically, if you take the first 109 Fibonacci numbers and divide each by 10 raised to the power of 109 MINUS its position in the Fibonacci sequence (including 0), the sum of those 109 numbers is $\frac{1}{109}$. (See **89**.)

110 $\left[\ 2 \times 5 \times 11 \qquad 5^2 + 6^2 + 7^2\ \right]$

TO J.R.R. Tolkien, 110 was "eleventy." To readers of *Scientific American*, it was the number of regions in this William McGregor drawing, used by Martin Gardner as an April Fool's prank in 1975. Gardner claimed that this pattern could not be colored using only four colors, and many believed him. Just one year later, however, Appel and Haken proved that four colors were sufficient for any map. (See **4**.)

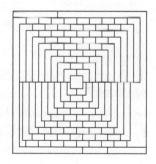

111 $\left[3 \times 37 \right]$

ANY 6 × 6 magic square has the property that the sum of any row, column, or diagonal equals 111. (The general formula for the magic constant of an $n \times n$ square is $\frac{(n^3 + n)}{2}$, and $111 = \frac{(6^3 + 6)}{2}$.) The pictured square dates back to the Middle Ages, when magic squares were accorded mystical properties. Speaking of sixes, 111 is the smallest number requiring six syllables in English, although the British style ("and" included) brings the total to seven.

6	32	3	34	35	1
7	11	27	28	8	30
19	14	16	15	23	24
18	20	22	21	17	13
25	29	10	9	26	12
36	5	33	4	2	31

112 $\left[2^4 \times 7 \right]$

WE saw in **21** that the smallest possible dissection of a square into squares with *distinct* integral sides requires that the large square be 112 units per side.

▼

IN 1945, the US Occupation Forces published "112 Gripes About the French" as a guide for troops stationed there. The title sounds like an effort to ridicule, but the actual purpose of the book was to promote cultural and historical understanding. As in:

6. We're always pulling the French out of a jam. Did they ever do anything for us?

The answer is yes, if you count the American Revolution, as in General Lafayette, 45,000 French army volunteers crossing the Atlantic in small boats, and over $6,000,000 in loans when a million dollars was a lot of money. And so on.

113 [prime]

$\frac{355}{113}$ is an extremely good approximation of π($\frac{355}{113}$ = 3.1415929 . . . , while π = 3.1415926 . . .). It was discovered in the fifth century AD by Chinese mathematician and astronomer Tsu Ch'ung-Chih.

114 [2 × 3 × 19]

THE number 114 apparently held special fascination for film director Stanley Kubrick, who brought us the CRM 114 radio in *Dr. Strangelove* and Serum 114, with which Alex was injected in *A Clockwork Orange*.

115 $\left[\,5 \times 23\,\right]$

THE Rule of 115 works just like the Rule of 72 (see **72**), except that it involves tripling instead of doubling. To find out how long it takes an investment to triple in value, divide the expected annual return into 115. For example, using the factorization of 115 above, an investment that returns 5% per year will triple in approximately 23 years. Just as the Rule of 72 works because 0.72 is close to the natural logarithm (base e) of 2, 1.15 is close to the natural logarithm of 3.

116 $\left[\,2^2 \times 29\,\right]$

THE Hundred Years' War between England and France was actually a series of conflicts between 1337 and 1453, a span of 116 years.

117 $\left[\,3^2 \times 13\,\right]$

THIS diagram shows a Heronian tetrahedron—a tetrahedron in which the sides (labeled in the diagram), faces, and volume are all rational numbers (fractions). The depicted figure is the *integral* Heronian tetrahedron whose largest side (117) is the smallest possible. (Yes, you read that right.) Its surface areas are 1170, 1800, 1890, and 2016 square units, and its volume is 18,144 cubic units.

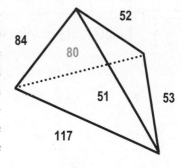

118 $\left[2 \times 59 \right]$

THE shortest ribbon length that solves the Christmas Package Problem with $n = 4$ is 118: Find four packages of different sizes but with equal length ribbons and equal volumes.

$$118 = 14 + 50 + 54 \qquad 14 \times 50 \times 54 = \mathbf{37,800}$$
$$118 = 15 + 40 + 63 \qquad 15 \times 40 \times 63 = \mathbf{37,800}$$
$$118 = 18 + 30 + 70 \qquad 18 \times 30 \times 70 = \mathbf{37,800}$$
$$118 = 21 + 25 + 72 \qquad 21 \times 25 \times 72 = \mathbf{37,800}$$

▼

A dime has 118 grooves on its side.

119 $\left[7 \times 17 \right]$

AND a quarter has 119 grooves on its side.

▼

A Pythagorean triple, the third smallest with consecutive legs, is 119–120–169.

(The smallest and best known is 3–4–5, and the second is 20–21–29.)

120 $\left[\, 2^3 \times 3 \times 5 \,\right]$

120 = 5! = 5 × 4 × 3 × 2 = the number of ways of arranging a five-card poker hand.

THE highest possible score for the first move in Scrabble using familiar words is 120: For *jukebox*, count a double letter score for X or J, plus a double word score and 50-point bonus. *Squeeze* and *quizzed* can also produce 120 points.

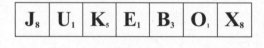

THE set {1, 3, 8, 120} has the unusual property that if you multiply any two of these numbers together and add 1, the result is a perfect square:

$1+1\times3$	$1+1\times8$	$1+1\times120$	$1+3\times8$	$1+3\times120$	$1+8\times120$
4	9	121	25	361	961

At each stage the number that is added to the sequence is the lowest possible. The next number in the sequence is 1680, followed by 23,408.

EACH of the interior angles of a hexagon measures 120 degrees.

121 $\left[\ 11^2 \qquad 3^0 + 3^1 + 3^2 + 3^3 + 3^4\ \right]$

THE number 121 is a palindrome and is also the square of a palindrome (11) and the square root of a palindrome (14641). There are infinitely many numbers of this type: Can you find the next one? (See Answers.)

▼

CHINESE checkers is neither Chinese nor checkers, but a standard board has 121 holes.

▼

IN math parlance, a number that is both a perfect square and a star number is called—no surprise here—a square star number. The next square star number after 121 is 11,881.

122 $\left[\ 2 \times 61\ \right]$

IN computing, a universally unique identifier, or UUID, is essentially a 128-bit construction of which 6 bits are claimed by version and variant, leaving 122 random bits. The total number of UUIDs is 2^{122}, a number with 37 digits.

Having a gigantic number of individual IDs is important because it makes the likelihood of an accidental repetition so small that it can be ignored.

123 $\left[\,3 \times 41\,\right]$

THE sum of the digits of 123 equals the product of the digits.

▼

START with any number, say, 829,432,154. Count up the number of even digits (5) and odd digits (4) and create a number using those two digits as well as their sum: 549. If you repeat the process starting with 549, you get 123. No matter what number you start with, you'll end up with 123 after some finite number of steps.

▼

IN the United Kingdom, dialing 123 gets you to British Telecom's "speaking clock," said to be accurate to within five thousandths of a second.

▼

THE number 123 is best known not as a number but for its digits, as in "easy as 1-2-3," "Lotus 1-2-3" (surely intended to be easier than, say, Visicalc), and Len Barry's "1-2-3," which never hit number 1 on the Billboard charts but reached number 2 in the United States and number 3 in the United Kingdom upon its release in 1965.

124 $\left[\, 2^2 \times 31 \,\right]$

THE United Kingdom has 124 postcode areas, defined as the first two letters in the postcode.

125 $\left[\, 5^3 \,\right]$

THE number 125 is a Friedman number, the term given to numbers that can be expressed as an equation using only the digits in the number itself: Behold the equation $125 = 5^{(1+2)}$.

126 $\left[\, 2 \times 3^2 \times 7 \,\right]$

126 is the biggest of the six magic numbers of physics, so called because atomic nuclei with 2, 8, 20, 50, 82, or 126 nucleons are especially stable. In 1963, Maria Goeppert-Mayer, Eugene Wigner, and J. Hans D. Jensen shared the Nobel Prize in Physics for their work in the "shell" method that sought to explain some of the higher magic numbers.

▼

THE equation $126 = {}_9C_5 = \frac{9!}{5!4!}$ is a bit of math shorthand, seen in many other places in the book, to represent the fact that there are 126 ways of choosing 5 (or 4) objects from an original set of 9. Now, frankly, a whole lot of numbers can be represented in this fashion by letting 9 and 5 equal, well, anything else. But this particular "choice" function has a real-life representation, because it indicates the number of possible 5–4 decisions from the Supreme Court of the United States.

127 $\left[\text{prime} \qquad -1 + 2^7\right]$

THERE are 127 matches required to determine the singles champion at Wimbledon—or any tournament with a full draw of 128. There are two ways to determine that 127 is the correct number. The first method begins by observing that the finals consists of one match, the semifinals two matches, and so on all the way back to the 64-match first round. The sum $1 + 2 + 4 + 8 + 16 + 32 + 64$ equals 127. (In general, the sum of the first n powers of two, including 2^0, equals one less than the *next* power of two.) The other way of counting the matches is to note that each match knocks out one participant. Only one person doesn't lose at all, so the total number of matches equals $128 - 1 = 127$. This shortcut applies to draws of any size: When the number of players or teams in a tournament is not a power of two, "byes" are given to ensure that the *second* round is a power of two.

128 $\left[2^7\right]$

AS mentioned in **127**, the first round of any major tennis championship consists of 128 players. In general, a tournament with 2^n players will have n rounds. Not only is 128 the number of entrants in a grand slam singles event in tennis, it is the largest number that cannot be expressed as the sum of three distinct squares.

129 $[\,3 \times 43\,]$

WHEREAS 128 cannot be written as the sum of three distinct squares, 129 can be expressed as the sum of three distinct squares in two different ways. $129 = 100 + 25 + 4 = 10^2 + 5^2 + 2^2$, and $129 = 64 + 49 + 16 = 8^2 + 7^2 + 4^2$. (The smallest number with this same property is 62.) Upon including the representations $64 + 64 + 1$ and $121 + 4 + 4$, in which two squares are repeated, there are altogether four representations of 129 as the sum of three squares, and 129 is the smallest number with four representations.

130 $[\,2 \times 5 \times 13\,]$

SPEAKING of sums of squares, the smallest divisors of 130 are 1, 2, 5, and 10, and $130 = 1^2 + 2^2 + 5^2 + 10^2$. No other number equals the sum of the squares of its first four divisors.

131 $[\,\text{prime}\,]$

THE number 131 is not only prime, it is a permutable prime, so called because the other numbers that can be obtained by permuting its digits, namely 113 and 311, are themselves prime. And, of course, 131 can be made by overlapping the primes 13 and 31.

132 $\left[2^2 \times 3 \times 11 \right]$

$$132 = 12 + 13 + 21 + 23 + 31 + 32$$

IN other words, 132 is the sum of all the two-digit numbers that can be formed using its own digits. In particular, 132 is the smallest such number. Care to guess what the next one is? (See Answers.)

▼

THE sixth Catalan number is 132. Catalan numbers show up in a wide variety of contexts in the field of combinatorics. There are 132 ways of dividing an octagon into six triangles. Depicted below are two of 132 ways in which six rectangles can cover the same "step diagram."

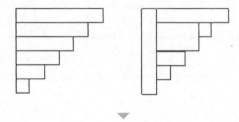

▼

THE general formula for the *n*th Catalan number is $\frac{2nCn}{(n + 1)}$, which equals $\frac{(2n)!}{(n!)^2(n + 1)}$.

133 $\left[7 \times 19 \right]$

ACCORDING to a seminal classification by D. E. Wilson and D. M. Reeder in 1993, there are 133 families of mammals. The names of mammal families are readily detected because they end in *-idae*, as in Caenolestidae (shrew and rat opossums), Dasypodidae (anteaters), Odobenidae (walrus), Erethizontidae (New World porcupines), and Castoridae (beavers).

Surprisingly, 19 of these 133 families, fully one-seventh, are some form of bat.

134 $\left[\, 2 \times 67 \,\right]$

USING Roman numerals, 134 is a Friedman number: $\text{CXXXIV} = \text{XV}^*(\frac{\text{XC}}{\text{X}}) - \text{I}$. (See **125**.)

135 $\left[\, 3^3 \times 5 \qquad 1^1 + 3^2 + 5^3 \,\right]$

EACH angle of a regular octagon is 135 degrees.

136 $\left[\, 2^3 \times 17 \,\right]$

ACCORDING to the standard Myers-Briggs personality classification scheme, there are 16 distinct personality types (see **16**). But if the psychotherapist who has to be familiar with all these types has it rough, the situation gets quite a bit worse for couples therapists. The number of ways you can choose two distinct types from a total of 16 equals $\frac{16(15)}{2}$, or 120. And if the two members of a couple are the same personality type, that's 16 more possibilities, for a grand total of 136.

▼

IF you sum the cubes of the digits of 136, you get $1^3 + 3^3 + 6^3 = 244$. If you repeat the process, you get $2^3 + 4^3 + 4^3 = 136$. The only other pair of numbers that produces this same symmetry is (919,1459).

137 [prime]

PHYSICIST Wolfgang Pauli is said to have died in hospital room 137, this after spending a lifetime trying to demonstrate that 137 is the "fine-structure constant."

138 [$2 \times 3 \times 23$]

UNTIL the punk band the Misfits came out with their song "We Are 138" in 1982, there was absolutely nothing to say about this number. In some sense there still isn't.

139 [prime]

A multiyear computer study directed by master puzzle designer Bill Cutler demonstrated that the maximum number of pieces into which a six-piece burr puzzle can be maneuvered without falling apart is 139. The pictured puzzle is one of those that produces the maximum.

140 $\left[2^2 \times 5 \times 7 \right]$

A "knight's tour" on a chessboard is created by placing a knight on any one of the 64 squares and traversing a path that takes the knight to each of the other squares once and only once. A magic knight's tour adds the provision that if you number each position of the knight as it makes its way around the board, the resulting set of 64 numbers forms a magic square. A semi-magic knight's tour (history's first one, constructed by William Beverley in 1848, is shown below) is a tour that creates a semi-magic square—one whose rows and columns sum to a magic constant (260) but whose diagonals do not. (In this case they sum to 280 and 212.) A pure 8×8 magic knight's tour was shown to be impossible in 2003 by J. C. Meyrignac and Guenter Sterntenbrink, whose research also revealed 140 different geometric forms for a semi-magic tour.

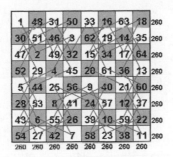

141 $\left[3 \times 47 \right]$

A Cullen number, named after Irish mathematician-turned-theologian Rev. James Cullen (1867–1933), is a number of the form $n \cdot 2^n + 1$. If $n = 1$ then $n \cdot 2^n + 1 = 3$, a prime number, but the next Cullen prime doesn't occur until $n = 141$. It is unknown whether an infinite number of Cullen primes exist.

142 $\left[\, 2 \times 71 \, \right]$

A pound equals 453.59 grams. An ounce equals $\frac{1}{16}$ of a pound. A carat equals 200 milligrams. Put it all together and you find out that an ounce is slightly less than 142 carats.

143 $\left[\, 11 \times 13 \, \right]$

THE number 143 is a factor of 1001 and therefore divides evenly into any number of the form abc,abc.

144 $\left[\, 2^4 \times 3^2 \qquad (1 + 4 + 4) \, (1 \times 4 \times 4) \, \right]$

JUST as a group of 12 is called a dozen, a group of 144, or a dozen dozen, is called a gross. In particular, 144 is 12 squared. As luck would have it, 144 is also the twelfth Fibonacci number. The only other square in the Fibonacci sequence is 1.

▼

THE number 144 also played a role in a counterexample to Euler's "Sum of Powers Conjecture." Euler had conjectured that for $n > 2$, at least n nth powers are required to add to a number that is itself an nth power. (Sort of a cousin of Fermat's Last Theorem.) This conjecture went unresolved until 1967, when L. J. Lander and T. R. Parkin discovered the equation $144^5 = 27^5 + 84^5 + 110^5 + 133^5$, meaning that a fifth power could be the sum of only four fifth powers.

145 $\left[\, 2^2 \times 5 \times 7 \,\right]$

$$145 = 1 + 24 + 120 = 1! + 4! + 5!$$

THE only other number (besides the trivial cases of 1 and 2) that equals the sum of the factorials of its digits is 40,585.

▼

NOW start with any positive integer and add up the squares of its digits. For example, if you choose 769, you get $7^2 + 6^2 + 9^2 = 166$. Do the same again and you get $1^2 + 6^2 + 6^2 = 73$. Next you get $7^2 + 3^2 = 58$. Remarkably, if you keep going, one of two things will happen: (1) You'll get to 1 and stay there forever; (2) You'll arrive at a loop of eight numbers, the largest of which is 145, and you'll stay in that loop forever. In particular, if you start with 769 you'll get to 145 after just five steps. The full loop involving 145 is {145, 42, 20, 4, 16, 37, 58, 89, 145}.

146 $\left[\, 2 \times 73 \,\right]$

$$146 = 1 + 4 + 9 + 16 + 25 + 36 + 25 + 16 + 9 + 4 + 1$$

THE above equation arises in calculating probabilities in the rolling of two pairs of dice. There is one way for the sums of both pairs to equal 2, four ways for the sums of both pairs to equal 3, and so on, with a "7" being the most common sum and with the other likelihoods being symmetric around 7.

▼

IN probabilistic terms, the equation means that there are 146 ways (out of 1,296) to roll two pairs of dice so that the sum of the rolls is the same. So the probability of a greater sum from the first pair of dice equals $\frac{575}{1,296}$, the probability of the second sum being greater is also $\frac{575}{1,296}$, while the probability that the two sums are equal is $\frac{146}{1,296}$.

147 [prime]

IN the absence of fouls, 147 is the highest possible score that can be achieved on a snooker break.

148 [$2^2 \times 37$]

A vampire number is a number whose digits can be regrouped into two smaller numbers that multiply to the original. There are 148 six-digit vampire numbers (assuming you can't add extras by padding with zeroes), the smallest being $102{,}510 = 201 \times 510$ and the largest being $939{,}658 = 953 \times 986$.

149 [prime]

ALTHOUGH 149 is prime, its main properties revolve around perfect squares, as follows:

149 is the sum of two perfect squares (100 and 49).
149 is the concatenation of two squares (1 and 49).
149 is also the sum of three consecutive squares ($6^2 + 7^2 + 8^2$).
149 is also the concatenation of three squares (1, 4, and 9).

150 $\left[\, 2 \times 3 \times 5^2 \,\right]$

AUSTRALIA'S House of Representatives has 150 members, each representing a different electoral division. A sesquicentennial is a 150th anniversary. The Bible contains precisely 150 Psalms. All well and good, but the most interesting application of the number 150 can be traced to British anthropologist Robin Dunbar, whose research indicated that 150 is the maximum number of people that can maintain a social relationship. Dunbar's research is actually a formula that provides a maximum group size per species as a function of the size of the neocortex of that species. The mean figure for *Homo sapiens* was 147.8, which Dunbar conveniently rounded to 150, known as Dunbar's number.

▼

AS Malcolm Gladwell notes in *The Tipping Point*, Dunbar's research confirmed the role of 150 (or at least 150ish) in such groups as the Hutterites, who had kept their colonies limited to 150 long before the existence of social psychology. Armies from Roman times to present day have kept units small to improve cohesion. And the manufacturers of Gore-Tex have specifically kept the number of employees at a given plant under 150 for the same reason, discovering Dunbar's number purely through experience.

151 $\left[\, \text{prime} \,\right]$

A prime palindrome. Also the number of Pokémon figures.

152 $\left[2^3 \times 19 \right]$

AN American mah-jongg set consists of 152 tiles: 108 suit tiles, 16 wind tiles, 12 dragon tiles, 8 flower tiles, and 8 jokers.

153 $\left[3^2 \times 17 \right]$

153 = $1^3 + 5^3 + 3^3$ and is the smallest of four integers with that same property (370, 371, and 407 are the only other numbers that equal the sum of the cubes of their digits, while *no number* is the sum of the squares of its digits). 153 also equals $1 + 2 + 3 + 4 + 5 + 6 + 7 + 8 + 9 + 10 + 11 + 12 + 13 + 14 + 15 + 16 + 17$, (making it the seventeenth triangular number), and equals $1! + 2! + 3! + 4! + 5!$ as well.

154 $\left[2 \times 7 \times 11 \right]$

BETWEEN 1904 and 1960 (with the exception of 1919), a baseball season consisted of 154 games: With eight teams in each league, each team played the other seven teams in its league 22 times apiece. (See **162**.)

154! + 1 (1 plus the product of the first 14 integers) is prime and for many years was the largest known prime number of that form. Primes of the form $n! \pm 1$ are known as *factorial primes*. It has been conjectured that there are infinitely many factorial primes. Note that if p is prime and $p < n$, then $n! + p$ can never be prime, because it is divisible by p.

155 $\left[5 \times 31 \right]$

AS we saw in **148**, any number whose digits can be rearranged to form two numbers that multiply to the original is called a *vampire number*. There are a total of 155 six-digit vampire numbers if you include the seven lousy ones with zeroes at the end, as in $150 \times 930 = 139{,}500$. In 2003, the 100-digit vampire number 97546105798506325258725803993761085200485109828763944370672506919920461931419704187863834796 31226428 was found. It equals 9876543210987654321098765432109876543210899077 6898 × 9876543210987654321098765432109976543211 0002523486.

156 $\left[2^2 \times 3 \times 13 \right]$

SUPPOSE a clock strikes only on the hour. Over a 12-hour period, the total number of strikes equals the sum of 1 through 12, otherwise known as the twelfth triangular number, or 78. Therefore, the total number of strikes in a full day is $2 \times 78 = 156$.

157 $\left[\text{prime} \right]$

$$157^2 = 24{,}649 \text{ and } 158^2 = 24{,}964$$

AT one time 157 was the largest known number whose square consists of the same digits as the square of its successor. Can you come up with the *smallest* pair of consecutive numbers whose squares use the same digits? And, for those who truly love a challenge, can you come up with a *bigger* pair of consecutive numbers with the same property? (See Answers.)

158 [2 × 79]

THE Greek national anthem is based on a 158-verse poem written by Dionysios Solomos. The poem "Hymn to the Freedom" was inspired by the Greek Revolution of 1821 against the Ottoman Empire. The anthem was officially adopted in 1864.

159 [3 × 53]

A barrel of oil contains 159 liters.

160 [2⁵ × 5]

$$160 \; [\, 2^5 \times 5 \,]$$

DON'T believe this poster. We'll set you straight a few pages from now.

161 $[\,7 \times 23\,]$

ALL primes other than 2 and 3 are of the form $6n \pm 1$. Where 161 fits in is that all numbers greater than 161 can be expressed as the sum of distinct primes specifically of the form $6n - 1$. For example,

$$162 = 47 + 41 + 29 + 23 + 17 + 5$$
$$163 = 101 + 29 + 17 + 11 + 5$$
$$164 = 131 + 17 + 11 + 5$$
$$165 = 89 + 71 + 5$$
$$166 = 107 + 59$$

And so on. There's no particular pattern at work here, but the construction is apparently always possible from 162 on.

162 $[\,2 \times 3^4\,]$

SINCE 1961, the number of regular-season games in Major League Baseball has been 162 (see **154**).

But there's an interesting and forgotten wrinkle here. The 154-game schedule in effect before 1961 made sense because there were eight teams in each league. Each team therefore had seven opponents, and because $154 = 7 \times 22$, each team could play every other team precisely 22 times.

So far so good, but when the American League expanded to 10 teams in 1961 with the addition of the Los Angeles Angels and Minnesota Twins, that even divisibility would have been lost, but for the move to a 162-game schedule. Yet the National League still had only eight teams, because the New York Mets and Houston Colt .45s didn't arrive until 1962. So which league gave up its nice, clean schedule?

The answer is neither. While AL teams played 18 games against each of nine opponents for a total of 162 games, the NL kept to its 154-game schedule for one final season. That's one reason why Roger Maris's extra eight games in which to break Babe Ruth's home run record were so conspicuous.

Nowadays, with extra divisions and interleague play, the notion of divisibility feels antiquated and the rumor (yes, it was just a rumor) of an asterisk on Maris's 61 home runs wouldn't get any traction. But things were different in 1961.

163 [prime]

THE number $e^{\pi\sqrt{163}}$ is very close to being an integer. Its value is 262,537,412,640,768,743.99999999999925.

This number was the subject of a 1965 April Fool's joke played on the readers of *Scientific American* by its famous columnist Martin Gardner. Gardner not only claimed that $e^{\pi\sqrt{163}}$ was integral, he credited Indian mathematician Ramanujan with having conjectured this "fact" in a 1914 paper, even though French mathematician Charles Hermite knew otherwise as far back as 1859. Ever since, the number $e^{\pi\sqrt{163}}$ has gone by the whimsical name of Ramanujan's constant.

164 [$2^2 \times 41$]

REMINISCENT of 149, 164 is the sum of two squares (100 and 64) and can be expressed as the concatenation of squares in two different ways: 1 and 64 or 16 and 4.

165 $[\ 3 \times 5 \times 11\ ]$

AS the sum of triangular numbers, 165 can be found along the third diagonal of Pascal's Triangle—in the eleventh row. Otherwise stated, the number of ways of choosing 3 objects from a set of 11 equals 165.

166 $[\ 2 \times 83\ ]$

AN example of a Smith number is 166—the sum of its digits ($1 + 6 + 6 = 13$) equals the sum of the digits of its prime factors ($166 = 2 \times 83$ and $2 + 8 + 3 = 13$). By convention, the prime factors are summed in accordance with their multiplicity; therefore 4 ($= 2 \times 2$ and $2 + 2$) is the first Smith number, while 166 is the eighth. The first 14 Smith numbers all have digital sums of 4, 9, or 13.

Smith numbers came into being in 1982, when Lehigh University mathematician Albert Wilansky pondered his brother-in-law's phone number: 493-7775. As a seven-digit number, this factors into $3 \times 5 \times 5 \times 65837$, and $4 + 9 + 3 + 7 + 7 + 7 + 5 = 42 = 3 + 5 + 5 + 6 + 5 + 8 + 3 + 7$. Wilansky was so amazed by his discovery that he named it after his brother-in-law, Harold Smith.

167 $[\ \text{prime}\ ]$

COAXIAL cable has a bandwidth from 0 to 1 GHz, and can therefore accommodate 167 separate TV signals at 6 MHz each.

MARTINA Navratilova won 167 singles titles, a record for the Open Era.

▼

APPARENTLY the poster at **160** was just an approximation, because an actual count came up with 167 bullet holes (entrance and exit) in Bonnie and Clyde's car following the May 23, 1934, ambush that claimed the bank robbers' lives. Moral of the story: Don't believe round numbers.

168 $\left[\, 2^3 \times 3 \times 7 \,\right]$

IF an activity is literally done 24/7 for a full week, it is done for $24 \times 7 = 168$ hours.

▼

THERE are 168 possible knight's moves going up the chessboard. The number in each square to the right gives the possible upward knight moves starting from that square.

0	0	0	0	0	0	0	0
1	1	2	2	2	2	1	1
2	3	4	4	4	4	3	2
2	3	4	4	4	4	3	2
2	3	4	4	4	4	3	2
2	3	4	4	4	4	3	2
2	3	4	4	4	4	3	2
2	3	4	4	4	4	3	2

169 [13^2]

WHILE $13 \times 13 = 169$, $31 \times 31 = 961$, the only such construction in which no number has repeated digits.

▼

THERE are 169 functionally distinct (two-card) starting hands in Texas hold 'em poker: 13 having the same rank, 78 of different ranks in the same suit, and 78 of different ranks in different suits.

170 [$2 \times 5 \times 17$]

THE Athenian trireme was powered by 170 oarsmen on three tiers: 31 on each side of the top level, and 27 per side on each of the lower two levels.

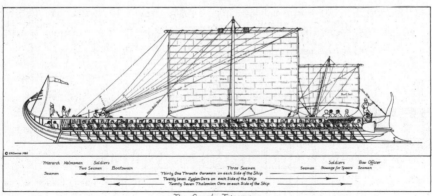

The Greek Trireme

171 $\left[\, 3^2 \times 19 \,\right]$

171 is triangular, being the sum of 1 through 18.

▼

HOW many Starbucks are there in Manhattan? The number changes, but on Friday, June 29, 2007, they totaled 171, and in a celebrated video, writer/comedian Mark Malkoff visited each and every one of them beginning that day at 5:30 a.m. and ending at 2:56 a.m. on Saturday the 30th.

172 $\left[\, 2^2 \times 43 \,\right]$

172 = $(4 \times 36) + (4 \times 6) + (4 \times 1)$, so 172 = 444 in base 6.

▼

EVERY year has between one and three Friday the 13ths, and there will be a total of 172 Friday the 13ths in the twenty-first century, starting with April 13, 2001, and ending with August 13, 2100.

173 $\left[\, \text{prime} \,\right]$

173 + 286 = 459 is the smallest sum that can be created using each of the nine nonzero digits precisely once. The largest sum is 981, which can be created in two ways: 324 + 657 = 981 and 235 + 746 = 981.

▼

THE period between lunar eclipses is roughly 173 days—slightly less than six lunar months.

174 $\left[\, 2 \times 3 \times 29 \,\right]$

BY regrouping the factorization of 174 in two different ways, we come up with the side-by-side equations $58 \times 3 = \mathbf{174} = 29 \times 6$, which together use each of the nine nonzero digits exactly once.

175 $\left[\, 5^2 \times 7 \qquad 100 + 1^2 + 7^2 + 5^2 \qquad 1^1 + 7^2 + 5^3 \,\right]$

THE sum of the first 49 positive integers equals $\frac{49(50)}{2} = 1{,}225$. Dividing by 7 yields 175, so each row, column, and diagonal of a 7×7 magic square must add up to 175. In a superstition that dates back at least to the sixteenth century, magic squares have been assigned to various planets in the solar system. Seven planets were known at that time, and the magic square below is known as the Venus magic square.

22	47	16	41	10	35	4
5	23	48	17	42	11	29
30	6	24	49	18	36	12
13	31	7	25	43	19	37
38	14	32	1	26	44	20
21	39	8	33	2	27	45
46	15	40	9	34	3	28

176 $\left[\, 2^4 \times 11 \,\right]$

177 $\left[\, 3 \times 59 \,\right]$

THE numbers 176 and 177 are linked by the diagram below, which creates a 176 × 177 rectangle—"almost" a square—by using using 11 smaller squares of distinct sizes. The minimum number of distinct squares needed to produce an actual square equals 21. (See **21**!)

Rectangles subdivided into squares have been used to model certain types of electrical networks. Within such a model, the sizes of the individual squares correspond to the currents and/or voltages of the network itself.

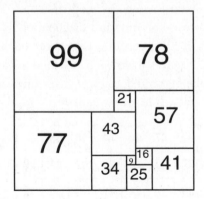

178 $\left[\, 2 \times 89 \,\right]$

THERE are 220 4 × 4 magic squares before rotations. Of these 220 squares, 178 are "balanced," meaning that every row, column, and diagonal has two numbers between 1 and 8 and two numbers between 9 and 16.

179 $\left[\, \text{prime} \qquad 17 \times 9 + 17 + 9 \,\right]$

THE equation $179 = 11 \times 15 + 14$ enables us to conclude that every year has 179 days whose day of the month is an even number. February has 14 even days whether or not it is a leap year.

180 $\left[\, 2^2 \times 3^2 \times 5 \qquad (10 - 1)(10 - 8)(10 - 0) \,\right]$

IF you take the product of the first seven positive integers and divide it by their sum, you get 180.

▼

IN most areas of the United States, 180 is the standard number of days in the school year. Among other things, students learn that a semicircle consists of 180 degrees, and that the sum of the degrees in the three angles of a triangle also equals 180 degrees. These two statements are related, as shown by the standard proof of the triangle result:

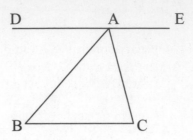

In the triangle ABC above, just draw a line through A that is parallel to the base BC. (Euclid's parallel postulate says that we can do this, and if it's okay by Euclid, it's okay by me.) From elementary geometry, the fact that DE and BC are parallel means that the angles ACB and EAC are equal, and likewise for DAB and ABC. Therefore, since the angle BAC is equal to itself, the sum of the angles in the triangle equals the number of degrees between points D and E, and that's obviously a half circle, or 180 degrees.

One corollary of this result is that the sum of the angles of a regular polygon with n sides equals 180 times $(n - 2)$, because such a figure can be divided into $(n - 2)$ triangles.

▼

BECAUSE 180 degrees is a half circle, *doing a 180* has come to mean "turning around," whether in a vehicle or in the sense of completely changing one's mind.

181 [prime]

ONCE major league baseball moved to a playoff system involving a five-game divisional playoff followed by a seven-game league championship series and, of course, a seven-game World Series, a team could theoretically play 162 + 5 + 7 + 7 = 181 games in a single season.

RECALL that the game of Go is played on a 19 × 19 grid. That gives 19^2, or 361 total intersections on which to place pieces. Since black goes first, there must be 181 black pieces and 180 white pieces to yield 361 in all.

182 $\left[\, 2 \times 7 \times 13 \,\right]$

THE number 182 made a brief, unwanted, and possibly apocryphal appearance in the construction of the Fahrenheit scale. According to one of many tales on the subject, Daniel Gabriel Fahrenheit (1686–1736) wanted the freezing points and boiling points of water to be separated by 180 degrees, as they are today, but he originally assigned 30 degrees as the freezing point of water, with 0 degrees the freezing point of a 50-50 salt/water combination. When, however, he measured the boiling point of water as 212°F, the undesirable 182-degree gap caused him to readjust the freezing point to 32°F.

183 $\left[\, 3 \times 61 \,\right]$

184 $\left[\, 2^3 \times 23 \,\right]$

THE numbers 183 and 184 are linked by more than just proximity. For starters, the number 183,184 is a perfect square (428^2), and is the smallest square obtained by concatenating two consecutive integers. (The only other six-digit numbers that work are $328,329 = 573^2$, $528,529 = 727^2$, and $715,716 = 846^2$.)

▼

THE two numbers also appear in the realm of "self-avoiding walks." Below, on the left, is one of 183 possible 7-step paths on a 7 × 7 grid such that the first step is to the right and no subsequent steps cross an existing path. Below, on the right, is one of 184 possible self-avoiding rook paths of order four, in which a rook goes from one corner of a 4 × 4 grid to the opposite corner, again without revisiting any spot along the tour.

Self-avoiding walks have been used in the study of polymers, solvents, and other chemical substances whose physical properties can be mimicked with a lattice-type structure.

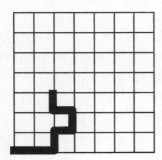 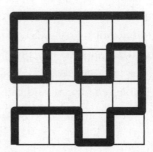

185 [5 × 37]

THE number 12,421 is the first of 185 five-digit mountain primes, so called because the digital values peak in the middle.

186 $\left[\, 2 \times 3 \times 31 \,\right]$

THERE are 186 days between the spring and fall equinoxes. Note that this number easily exceeds half a year. The difference between those 186 days and the 179 days between the fall and spring equinoxes is explained by the fact that the Earth maintains an elliptical orbit around the sun, so the distance traveled between spring and fall is actually greater. Not only that, the Earth moves slightly faster when closer to the sun, and the closest point, the perihelion, comes in early January, close to the Winter Solstice. This is a special case of Kepler's Second Law, which states that the radius vector from the sun to the earth sweeps out equal areas in equal times.

▼

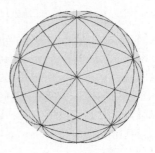

SPEAKING of spheres, the diagram shows a so-called icosahedral tiling of the sphere by triangles. This particular tiling is called a (2,3,5) tiling, for which the angles of each triangle are $\frac{180}{2}$, $\frac{180}{3}$, and $\frac{180}{5}$ degrees. Add those up and you get $90 + 60 + 36 = 186$ degrees, a reminder that the famous rule we proved in 180 that the sum of the angles of a triangle equals 180 degrees, well, that rule only applied in two dimensions.

187 $\left[\, 11 \times 17 \,\right]$

AS an extension of the birthday paradox (see **23**), if you have 187 people in a room (not necessarily a good idea, but let's just suppose), the odds are greater than 50% that *four* of the people in the room share a birthday.

188 $\left[\, 2^2 \times 47 \,\right]$

THE number $188 = 1 + 4 + 9 + 25 + 49 + 100$, the sum of six distinct squares. All numbers > 188 can be expressed as the sum of at most *five* distinct squares.

189 $\left[\, 3^3 \times 7 \,\right]$

$$189 = 12 + 34 + 56 + 78 + 9$$

▼

BY some counts, the English language contains 189 irregular verbs, starting with *abide* and ending with *write*.

▼

ELSEWHERE in language, the Braille alphabet (Braille II) contains a total of 189 one-cell and two-cell contractions.

190 $\left[\, 2 \times 5 \times 19 \,\right]$

IF you write out the factorization of 190 in Roman numerals, you get II × V × XIX. Each of these distinct prime factors is a palindrome, as is the product, CXC, and no number bigger than 190 has this property.

▼

JUST as X's play a vital role in Roman numerals, they play a vital and especially desirable role in bowling, where an X denotes a strike. But if your X's are hard to come by, perhaps you should know that 190 is the highest score that can be achieved without rolling a single strike.

191 [prime]

IF you had one of every denomination of coin produced by the US Mint, you'd have a silver dollar, a half-dollar, a quarter, a dime, a nickel, and a penny, for a total of 100 + 50 + 25 + 10 + 5 + 1 = 191 cents.

192 [$2^6 \times 3$]

NOTE that the factorization of 192 involves a bunch of 2's and a single 3. Numbers such as these have a lot of divisors, and in fact 192 is the smallest number with 14 divisors: 1, 2, 3, 4, 6, 8, 12, 16, 24, 32, 48, 64, 96, and 192 itself.

TAKE two sticks 20 inches in length and join them at one end. If you move the other ends until they are 24 inches apart, you will have created a triangle with an area of 192 square inches. If you keep spreading the sticks, the area will get bigger to a point, then fall. By the time the ends of the sticks are exactly 32 inches apart, the triangle they create will again be precisely 192 square inches. Somewhere in between a maximum area is obtained: Care to guess what that maximum is? (See Answers.)

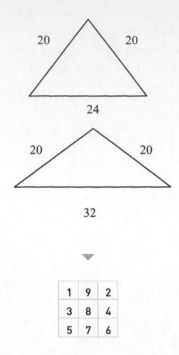

THE above 3 × 3 square is nothing more than the first three multiples of 192 stacked upon one another. What makes it special, of course, is that it uses each of the digits 1 through 9 precisely once.

193 [prime]

IN the words of zoologist Desmond Morris, "There are 193 species of monkeys and apes, 192 of them are covered with hair. The exception is a naked ape self-named Homo sapiens." Morris wrote about that exception in his revolutionary 1967 book, *The Naked Ape*.

194 $[\,2 \times 97\,]$

THE Roman Catholic Church has 194 dioceses within the United States, a number that becomes 195 upon the inclusion of the Archdiocese of the Military Services.

195 $[\,3 \times 5 \times 13\,]$

STEINWAY'S "Peace Piano," crafted in 2004, contained the flags of 195 nations around its perimeter, representing the number of countries in the United Nations at that time.

196 $[\,2^2 \times 7^2\,]$

MOST numbers become a palindrome by repeatedly reversing their digits and adding: For example, a starting number of 349 yields 349 + 943 = 1292, 1292 + 2921 = 4213, and 4213 + 3124 = 7337. The process isn't always this swift: Starting with 89 takes 24 steps before you arrive at a palindrome. Where does 196 fit in? Astonishingly, it is not known whether you *ever* get a palindrome starting with 196. It is the smallest number whose conversion is uncertain. Numbers that aren't known to ever convert are called Lychrel numbers, an amusing class of numbers in that by the nature of the definition of a Lychrel number, it's hard to be certain that a particular number is one!

197 [prime]

197 belongs to a restrictive club called the Keith numbers. The first Keith number is 14, as follows: If you begin a sequence with 1, 4 (in other words, the digits of 14), after which each new member of the sequence is obtained by adding the previous two (as in the Fibonacci numbers), the sequence proceeds 1, 4, 5, 9, **14**, reaching the starting number. With 197, the sequence begins with 1, 9, 7, and then adds the previous *three* numbers, forming 1, 9, 7, 17, 33, 57, 107, **197**. Keith numbers are extremely rare: Last we checked, the entire list of known Keith numbers consisted of only 95 numbers, from 14 all the way to the 29-digit whopper 70,267,375,510,207,885,242,218,837,404.

198 [$2 \times 3^2 \times 11$ $(1 + 9 + 8) \times 11$ $11 + 99 + 88$]

THERE are 198 palindromes under 10,000, as follows:

```
  9 – one digit:  1, 2, ... 9
  9 – two digits:  11, 22, ... 99
 90 – three digits: 101, 111, ... 191, 202, 212, ... 292, ... 909, 919, ... 999
+ 90 – four digits: 1001, 1111, 1991, 2002, 2112, ... 2992, ... 9009, 9119, ... 9999
198
```

199 [prime]

THE number 199 is a permutable prime, meaning that it remains prime even when you rearrange its digits.

BECAUSE 199 has a repeated digit, all that means is that the numbers 919 and 991 are prime as well. On the other hand, 199 offers an extra flourish that it becomes 661 when viewed upside down, and that number is prime as well.

▼

NOTE that 191, 193, 197, and 199 are all prime.

200 $\left[\, 2^3 \times 5^2 \,\right]$

OUR final entry, 200, has a little bit of everything, sort of like this book. First, we have a touch of mathematics:

In sharp contrast to 199, 200 is the smallest unprimable number: Not only is 200 composite, it remains composite if you change any one of its digits to any other number. Equivalently, the numbers 200, 201, 202, 203, 204, 205, 206, 207, 208, and 209 are all composite, the first such ten-number sequence.

There are $2^8 = 256$ subsets of the numbers $\{1, 2, 3, 4, 5, 6, 7, 8\}$. But only 200 of them are "weakly triple-free," meaning that they don't contain either $\{1, 2, 3\}$ or $\{2, 4, 6\}$ as a subset.

▼

AND then we have some sports:

In bowling, if you alternate strikes and spares for an entire string, you will have attained a score of 200, sometimes called the Dutch 200. Even more surprising, fighters who weigh more than 200 pounds are considered heavyweights.

▼

FINALLY, we have some approximations and lame tie-ins:

The human field of vision is said to be approximately 200 degrees. And there are approximately 200 seeds on a strawberry (just as there were 200 pounds on Darryl Strawberry when he entered the major leagues—however, no one was calling him a heavyweight, even though he certainly wasn't weakly triple-free).

If only this were Monopoly, you could now pass Go . . . and collect your $200.

ANSWERS

3 ▶

The ratio of odd numbers to even numbers in Pascal's Triangle approaches zero as the number of rows heads to infinity.

4 ▶

The proof that any number eventually gets to four after counting the letters in its English name (creating a new number, and continuing in this fashion) is far easier than you'd think. The first step is to notice that any number less than 4 has more letters than itself: ONE has 3 letters, TWO has 3, and THREE has 5. FOUR has 4 letters, and it is not hard to see that the number of letters in any number greater than FOUR is less than the number itself.

Starting with ONE and applying the letter-counting sequence produces the chain ONE-THREE-FIVE-FOUR. Starting with two produces TWO-THREE-FIVE-FOUR. Starting with THREE produces THREE-FIVE-FOUR. Now suppose you pick any number whatsoever. That's right, any number. If you count the letters in its English representation, you get a smaller number than you started with, so if you keep going, you'll eventually get to 1, 2, 3, or 4. But we've already seen that those numbers end up at FOUR, so we're done.

10 ▶

The letters in question are the last letters of the first ten positive integers.

The logarithm equation is actually easier than you think. The left hand side reduces to $\log_2(\log_9^{(\frac{1}{2^n})} 9)$, or $\log_2(2^n)$, which by definition equals n.

12 ▶

Suppose that a Buckyball has P pentagons and H hexagons. The total number of faces on the Buckyball is therefore $P + H$. The total number of edges is $\frac{(5P + 6H)}{2}$, the division by 2 accounting for the fact that every line in the Buckyball is a side of two figures. Similarly, the total number of vertices is $\frac{(5P + 6H)}{3}$. According to Euler's formula:

$$2 + \frac{(5P + 6H)}{2} = \frac{(5P + 6H)}{3} + (P + H)$$

Multiplying both sides by 6 yields:

$$12 + 3(5P + 6H) = 2(5P + 6H) + 6(P + H), \text{ which simplifies to}$$
$$12 + 15P + 18H = 16P + 18H$$

Miraculously, the H terms cancel and we are left with $P = 12$.

16 ▶

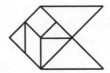

17 ▶

Here's the solution for the 17-clue Sudoku puzzle:

9	1	4	6	5	3	8	7	2
5	3	6	8	2	7	1	4	9
8	2	7	9	4	1	6	5	3
7	6	8	3	1	5	9	2	4
1	5	3	4	9	2	7	6	8
2	4	9	7	6	8	3	1	5
3	7	5	1	8	4	2	9	6
6	8	2	5	7	9	4	3	1
4	9	1	2	3	6	5	8	7

19 ▶

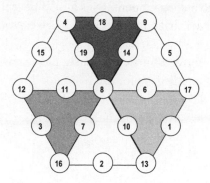

23 ▶

The 23-letter sequence gives you the alphabet, in order of appearance within the string ONE, TWO, THREE, FOUR, FIVE, and so on. The letter C doesn't appear until ONE OCTILLION, and J, K, and Z never appear at all.

24 ▶

The woodcut is from the second edition of *Geoffrey Chaucer's Canterbury Tales*, printed by William Caxton in 1483.

27 ▶

The number 15 has the same property, since $1 + 2 + 3 + 4 + 5 = 15$.

29 ▶

A rectangular block with sides of length a, b, and c will have volume equal to abc, a number with at least three prime factors (possibly repeated). But the combined volume of the 29 pentacubes is by definition 5×29, a number with only two prime factors. So, no matter how you combine the 29 pentacubes, they'll never form a perfect block.

31 ▶

At first glance this puzzle looks easy. Starting at the top, you subtract 72 from 99 to get 27. Then you subtract 27 from 45 to get 18, 18 from 39 to get 21, and so on. Placing 15 in the question mark continues the pattern perfectly, because $36 - 21 = 15$ and $28 - 15 = 13$. But this rule fails its final test at the bottom, when, agonizingly, $21 - 13$ gives you 8, not the 7 in the lowest circle.

Once you poke around a bit more you should have no trouble discovering that the number in any given circle is obtained by adding the individual digits in the two circles pointing toward it. Thus $7 + 2 + 9 + 9$ gives you 27, all the way to $1 + 3 + 2 + 1 = 7$ at the bottom. Along the way you see that replacing the question mark by $2 + 1 + 3 + 6 = 12$ works out just fine. So your answer is 12.

35 ▶

The diagram below is a 1930 creation of T. R. Dawson and demonstrates one way to move a knight 35 times without having it cross its own path.

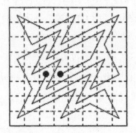

37 ▶

1. The number of hairs on a human head varies with hair color (blondes have more hair than brunettes, for example), but the upper limit is somewhere around 140,000. By contrast, more than 12 million people live in Tokyo. You can't put those 12 million people into 140,000 slots without at least two sharing a slot, and those two people by definition have the same number of hairs on their heads. Using a city as populous as Tokyo was, of course, overkill: The puzzle would have worked equally well using Stoke-on-Trent, England, or Huntsville, Alabama.

2. Divide the original equilateral triangle into four smaller equilateral triangles, as in the diagram. Each of the smaller triangles measures one inch on a side; in particular, any two points inside any of these triangles must be within an inch of one another. But if you had five points to place in these four triangles, the pigeonhole principle guarantees that (at least) two of the points must reside in the same triangle.

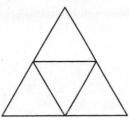

3. Choose any 10 numbers from the first 100 positive integers. The number of subsets of these 10 numbers is $2^{10} = 1,024$ (see the discussion in **4**). But the highest possible sum of 10 numbers in the range 1 through 100 equals $91 + 92 + 93 + \ldots + 100 = 955$. If there are more subsets than possible sums, at least two subsets must have the same sum. If those two sets are disjoint, you're done, and if not, just get rid of the common elements and the disjoint subsets that arise must also have the same sum, because all you've done is removed the repetitions.

39 ▶

There is one number smaller than 39 such that the sum of the primes in between the smallest and largest divisor equals the number itself. That number is 10: Its prime divisors are 2 and 5, and $2 + 3 + 5 = 10$.

48 ▶

A 5 × 12 rectangle also works. The interior uses 3 × 10 = 30 tiles and the border uses the remaining 30.

The 51 nations listed were the original 51 members of the United Nations in 1945.

The number 419 (= $60 \times 7 - 1$) leaves a remainder of $n - 1$ when divided by any of $n = 2, 3, 4, 5, 6,$ and 7.

This puzzle is often given using a chessboard, from which you extract diagonally opposite squares and ask if it can be covered using pieces that look like this:

The coloring of the squares makes it easier to reach the correct conclusion, which is no. Each of the 1×2 "dominoes" consists of a white square plus a black square, whereas the 62-square board obtained by removing two squares from the same diagonal must have 32 squares of one color (in this case black) and 30 squares of the other color. So no covering is possible.

69 ▶

The other number that equals the alphanumeric value of its Roman numeral representation is 63. (63 = LXIII = 12 + 24 + 9 + 9 + 9 = 63.)

72 ▶

List all possible sets of three positive integers whose product is 72. Here is the list, with the sum of the three numbers provided to the right of the equal sign:

1, 1, 72 = 74 ; 1, 2, 36 = 39 ; 1, 3, 24 = 28 ; 1, 4, 18 = 23 ; 1, 6, 12 = 19 ; 1, 8, 9 = 18 ; 2, 2, 18 = 22 ; 2, 3, 12 = 17 ; 2, 4, 9 = 15 ; 2, 6, 6 = 14 ; 3, 3, 8 = 14 ; 3, 4, 6 = 13

Had the number on the front door been anything but 14, there would be no ambiguity, so we can assume that the kids' ages are either 2, 6, and 6 or 3, 3, and 8. That's where the line "my youngest likes ice cream" comes in. Electing not to split hairs with the timing of the births of a set of twins, we conclude that the youngest is 2 and the other two kids each 6 years of age.

The next number after 72 that could work in this problem is 225, which can be represented as either $1 \times 15 \times 15$ or $3 \times 3 \times 25$, and both sets of three numbers add up to 31.

75 ▶

The missing possibilities are the six cases in which there are *two* ties:

[AB][CD]   [AC][BD]   [AD][BC]   [BC][AD]   [BD][AC]   [CD][AB]

79 ▶

The idea of the Coconut Problem is to work backward. Let x = the number of coconuts each thief took when they divvied up in the morning. Then the coconuts on hand after the third thief took his "share" (after giving one to the monkey) were $3x + 1$. The third thief took one-half of this number, so the amount left by the second thief was $(\frac{3}{2})(3x + 1) + 1$, which only makes sense if $3x + 1$ is an even number. Therefore x is odd. Okay, we've at least established something.

When you investigate the next level, you find that the conditions of the problem force $\frac{(3x+1)}{2}$ to be odd as well. And, whatever that number is, if you multiply by 3, add 1, and divide by 2, that number has to be odd, too.

This is getting confusing. Let's look at an actual example. If $x = 3$ there were 10 coconuts left in the morning, and therefore 11 prior to the third thief giving one to the monkey. That means the third thief took 5 coconuts (3 times 3, plus 1, divided by 2, as expected), so he must have encountered 16 originally (taking 5, leaving 10, and giving one to the monkey). That means that the second thief left 17 before giving one to the monkey, so he must have taken 8 ($\frac{(3\times5+1)}{2}$). But 8 is an even number, so we can't go up another level.

Similarly, if $x = 5$ we reach 8 even quicker, so that doesn't work. But if the final distribution was 7 coconuts apiece, the third thief must have taken 11 from a batch of 34, the second thief took 17 from a batch of 52, and the first thief took 26 from the original batch of 79. Since 26 is even, you can't go up another level, but you don't have to, as there were only three thieves.

81 ▶

It is possible to rewrite the expression as $(p + 1 + \frac{1}{p})(q + 1 + \frac{1}{q})(r + 1 + \frac{1}{r})$ $(s + 1 + \frac{1}{s})$.

A little thought will reveal that each of the four bracketed expressions must be greater than 3, so the product must exceed $3^4 = 81$.

83 ▶

The 83 numbers listed are the five-digit numbers whose squares contain nine of the ten digits. The numbers to the left of each batch of numbers indicate the digit that does *not* appear in the corresponding squares.

For the second question on the page, if you write out the first 500,000,000 positive integers, you will use precisely 500,000,000 1's. There are a total of 83 positive numbers n such that writing out 1 through n uses precisely n 1's, with 500000000 being right in the middle.

92 ▶

Five queens are sufficient to attack any square on a chessboard. Four queens aren't enough.

93 ▶

The 93 listed numbers form the complete list of five-digit palindromic primes. Those in boldface are the five-digit sequences that at this writing are working ZIP codes in the United States.

97 ▶

Hope the wording of the problem gave you a hint, because the answer is a cute but cheesy "NEGATIVE NINETY-SEVEN."

98 ▶

At the outset, you have 100 pounds of potatoes, of which 99% is water, so basically you have 99 pounds of water and 1 pound of something solid. Because the solid part is assumed not to change, it becomes 2% of the whole when the water weighs 49 pounds. At that point the total weight of the potatoes is 1 + 49 = 50 pounds.

104 ▶

Voilà: Twelve identical matchsticks arranged so that precisely three meet at each vertex.

108 ▶

12:00 and 11:20 are the other two times of day in which the hour figure multiplied by the minutes figure equals the number of degrees between the two hands, but for 11:20 you have to take the long way around.

121 ▶

There's a bit of a trick to this one. The next number that is both the square and square root of a palindrome is 10201, which is 101 squared and the square root of 104060401.

132 ▶

The next number that equals the sum of all the two-digit numbers that can be formed from its digits is 264, which of course is two times 132. From that you should be able to figure out the third number with this same property!

157 ▶

The smallest pair of consecutive numbers whose squares use the same digits is the pair $\{13,14\}$, as $13^2 = 169$ and $14^2 = 196$.
The next pair to have this property is $\{157,158\}$, and the one after that is $\{913,914\}$, where we have $913^2 = 833,569$ and $914^2 = 835,396$.

192 ▶

The maximum area is obtained when the legs form a right triangle, at which point the area is 200 square inches. The result relies on the fact that the area increases to a point, then decreases, and that maximal point represents some sort of symmetry.

ACKNOWLEDGMENTS

THE cover and title page suggest that I wrote this book by myself, but don't think for an instant that I didn't have massive amounts of help along the way. Fortunately, I now have the space to acknowledge all those individuals who gave me a boost. Here goes:

Let me start by thanking my agents, Jennifer Griffin and Mary Clemmey, who represented the book in New York and London, respectively. I'm sure that this particular book wasn't the easiest task of their careers, but they never gave up on it, for which I'm grateful. My editors in London, first Caroline MacArthur and then Mary Morris, inherited this project and were able to manage it nicely from afar. Nick Webb, who was the first to sign on to the book and the one who put in countless hours tracking its intricacies, was a delight to work with. Finally, Marian Lizzi at Perigee was there to put everything together, ably assisted by Christina Lundy, Tiffany Estreicher, and many others who worked just outside my radar screen.

Lynne Emmons of Arness House saved the day with her conversions of web-based artwork to formats suitable for reproduction on paper.

I am grateful for the conversations I had with first-rate mathematicians whose specialties overlapped with some of my goals for the manuscript. I would especially like to thank Norton Starr, David Kelly, Ross Honsberger, Richard Stanley, Arthur Benjamin, Raymond Smullyan, Gordon Prichett, and Herbert Scarf for their accessibility and their time.

I was able to use dozens of images that I first located on the web, each with a helpful creator or web designer who gave me the requisite permission. Thanks to Ken Knowlton at www.knowltonmosaics.com, Professor Carsten

Thomassen for his Thomassen graph, Bill Harrah for his beautiful Lincoln Memorial drawing, Stephane Gires and Mathilde Spriet at Gigamic for their wonderful games, and likewise for Kate Jones at Kadon Enterprises. The list goes on: Professor Heiko Harborth, with the help of Jens P. Bode, provided his famous toothpick drawing; Galen Frysinger, his Pont du Gard artistry; Achim Flammenkamp, his remarkable 52-square creation; Jim Loy, his 17-gon representation; Michel Emery for the wonderful kissing number diagram in L'Ouvert; Michel Guntern for the map of France at www.1800-Countries.com; Thomas Green at Contra Costa College for his bridges of Konigsberg diagram; Lawrence Charters for his calendars; Ed Rosenberg for his flags; A. Chatterjee in Mumbai for his chess diagram software; Rebecca Clark-Smith for her Patolli image; Allen Broughton at the Rose-Hulman Institute of Technology for his sphere-tiling drawing; Bill Cutler for the images of his clever work with Burr puzzles; Dan Thomasson for his knight's tour diagrams; and Andrew Ruddle for his pictures of triremes.

I'd also like to thank those people whose images, alas, ended up on the cutting room floor. There were many more of these than I'd like, including Laura Pecci at Winning Moves; Stefan Goya at www.psychicteddybear.com; Kristin MacQuarrie at www.harpconnection.com; Christopher Monckton, Alex Selby, and Oliver Riordan for their extraordinary Eternity puzzle images; Danny Smythe for his coconut picture; Theodor Lauppert for his images of Ishido; David Phillips for his graphs on the Baskerville effect; Steve and Tim Sommars for their clever work with octagons; Marc Gilbert for his website's Yankee Stadium image; Dollie at Jerry Ohlinger's Movie Material Store for pulling out a few real oldies; David DeJean for his colorful Lotus palette; Michael Kroeger and Thomas Detrie for their drawings of the icosahedron; William Waite for his Knit Pagoda puzzle; Don Hodges for his Pac-Man images; and, finally, Kristin Hylek at McDonald's.

There were also several organizations that chipped in to the effort. Thanks to Heidi Dettinger at Steinway & Sons, Sarah Hart and Jessica Zadlo at ThinkFun, Chris Holmes and Peter Costa at Owl Engineering, and Andrea Phillips at Fotosearch. I contacted several museums and other such collections during my image hunt, and was greatly assisted by Amanda

Turner and Helen Statham at the Ashmolean Museum at Oxford, Calune Eustache at United Media, Catherine Howell at the Victoria and Albert Museum, and Meghan Mazella and Valentina at the British Museum. And where I would have been without Brian Blankenburg at Getty Images is just too sad to contemplate.

Individuals who rose to the occasion included my sister, Eliza Miller, who tracked down a fabulous picture of Balanchine's *Serenade* from the Maine State Ballet. Jerry Slocum generously provided an image of the original 15 puzzle from 1880. Pete Malaspina provided access to all sorts of academic references that would otherwise have eluded me. Joseph West at Abaris Books gave me some helpful counsel. Boots Hinton gave me more information about Bonnie and Clyde than I ever thought I'd know. I regretted that I didn't have more than a couple of paragraphs to allot to those infamous gangsters, who died just five months after Boots was born—with Boots's father a member of the posse. And if there were any details that eluded Boots, Frank Ballinger was there as backup.

Finally, I benefited from a variety of mathematically oriented websites: MathWorld was absolutely indispensable; www.primecurios.com was delightfully bizarre; Mudd Math Fun Facts had some nice puzzles and curiosities; Erich Friedman at Stetson University is a one-man mathematical show whose work I greatly admire; Kevin Brown's work at www.math pages.com was helpful on several occasions; Ed Pegg's www.mathpuzzles.com was a delight to return to; and finally, moving overseas, the French website www.pagesperso-orange.fr/yoda.guillaume was a pleasure to translate through, and I have Gerard Villemin et al. to thank for that.

There. I think I've demonstrated that I didn't write this book alone. Thanks again to all.

PHOTO CREDITS

page 171: Braille Mosaic—Ken Knowlton, www.knowltonmosaics.com

page 180: Gordian's Knot—ThinkFun, Inc.

page 185: 72-sided sphere—Pacioli, Luca. *De divina proportione* (English: *On the Divine Proportion*), Luca Paganinem de Paganinus de Brescia

page 185: Archery Target—Wikimedia Commons/Alberto Barbati

page 187: Water Clock—Time Life Pictures/Getty Images

page 188: Stacked Balls—Wikimedia Commons/Greg L

page 199: Dart Board—www.CKSinfo.com

page 202: How to Tie a Tie—Yong Mao

page 209: Sunflower—John Foxx/Stockbyte/Getty Images

page 210: Baseball Diamond—Fotosearch

page 210: Dried Mud—Mike Norman/National Geographic/Getty Images

page 212: Abacus—Red Chopsticks/Getty Images

page 218: Builder's Old Measurement—Wikipedia/L'enciclopedia libera/Edoardo

page 220: Bell Curve—Chris Holmes/Owl Engineering

page 230: Matchstick Graph—Heiko Harborth

page 238: Poker Hand—Poker Pundit

page 246: Wooden Burr—Bill Cutler

page 247: Knight's Tour—Dan Thomasson

page 254: Bonnie & Clyde Poster—Frank Ballinger

page 259: Ship—Andrew Ruddle, The Trireme Trust

page 267: Sphere—Allen Broughton

ABOUT THE AUTHOR

Derrick Niederman is a mathematician by training but has been interested in puzzles and mathematical recreations since childhood. He is the author of several volumes of math puzzle books and two volumes of short mystery stories based on his work as "Inspector Forsooth," an early venture on America Online. He has produced more than 20 crosswords for the *Sunday New York Times*, and designed the mathematical puzzle 36 Cube based on research for this book. An investment writer by trade, Niederman received his PhD in mathematics from MIT in 1981 and continues to live in the Boston area.